Biotechnology Management

Biotechnology Management

by

SANGITA MALVEE
M.Sc. (Biochemistry)

BIOTECH

2008

Biotech Books

Delhi - 110 035

ISBN10 81-7622-192-9
ISBN13 978-81-7622-192-4

Published by : **BIOTECH BOOKS**
1123/74, Tri Nagar,
DELHI – 110 035
Phone: 27383999
e-mail: biotechbooks@yahoo.co.in

Showroom : 4762-63/23, Ansari Road, Darya Ganj,
NEW DELHI - 110 002
Phone: 23245578, 23244987

Laser Typesetting : **Classic Computer Services**
Delhi - 110 035

Printed at : **Chawla Offset Printers**
New Delhi – 110 052

PRINTED IN INDIA

Dedicated to

My Mother Smt. Vasundhara Kamik
My Son Master Sarang
My Daughter Shriya

ACKNOWLEDGMENTS

The list is very long, I want to thank all my well-wishers who support me and encourage me during this work. My special thanks to Mrs. Neena and Mr. Uday Pradhan, Mr. Piyush Bhatnagar and Mr Sandeep Yadav for their helpful suggestions, constant moral support and encouragement during the completion of this book.

Last but not least, I must thank Mr. Sunil Mittal, and staff of Biotech Books for their support.

Enormous efforts have been taken to avoid errors and mistakes, however there presence cannot be ruled out. Any shortcomings and lapses are entirely mine. Any suggestions to improve the standard of this book, indication towards errors, omissions and mistakes will be highly appreciated.

SANGITA

PREFACE

Biotechnology is technology that uses living organism or parts of organism to make or modify products, to improve quality of plant to develop animals or microorganisms for unambiguous use. Viable biotechnology consists of emergent range of consistent techniques, procedures and processes for practical applications in the health care, ·agriculture and industrial use commercial biotechnology includes research to services. If properly managed, it is widely recognized that biotechnology can play an essential role in fostering the economic and social development of developing countries. In the area of health care many biotechnologies products such as insulin, diagnostics, and vaccines have already been placed on the market and products such as recombinant resultant hepatitis B vaccine have gained pervasive international use in agriculture products such as diagnostics bio pesticides and bovine growth hormone have been in commercial use. Other products and technologies include improved seeds, new vaccines, novel food ingredients, biotechnology based techniques for the rapid detection and identification of toxic materials and several bioprocess technologies. The transfer and development-II of biotechnology requires large capital inputs, in case of many developing countries, which is not readily available. It require suitable infrastructure to absorb and to develop technology and to manage it properly and systematically. So to choose and adapt biotechnology can serve as a measurement of sustainable self-reliance that will allow it to participate fully in worldwide efforts to achieve sustainable development. Therefore it is necessary to study how one can develop Biotech Industry. This book primarily discusses overview of Biotechnology Industry, how one can develop Biotech Industry, what are primary factors that have to be taken care of. As an example Biopharmaceutical industry it's structure organization management is discussed. Some legal issue regarding human rights, animal rights *i.e.* Bioethics is also discussed. This book as a whole summarizes many of the important aspects of Biotechnology Management.

SANGITA

CONTENTS

Chapter 1

INTRODUCTION

Biotechnology is technology that uses living organism or parts of organism to make or modify products, to improve quality of plant to develop animals or microorganisms for unambiguous use. Viable biotechnology consists of emergent range of consistent techniques, procedures and processes for practical applications in the health care, agriculture and industrial use commercial biotechnology includes research to services.

If properly managed, it is widely recognized that biotechnology can play an essential role in fostering the economic and social development of developing countries in the area of health care many biotechnologies products such as insulin, diagnostics, and vaccines have already been placed on the market and products such as recombinant resultant hepatitis B vaccine have gained pervasive international use in agriculture products such as diagnostics bio pesticides and bovine growth hormone have been in commercial use. Other products and technologies include improved seeds, new vaccines, novel food ingredients, biotechnology based techniques for the rapid detection and identification of toxic materials and several bioprocessing technologies. Biotechnological inventions are based on new scientific discoveries it offers novel approaches for striking a balance between development needs and environmental preservation. The transfer and development of biotechnology requires large capital inputs, in case of many developing countries, which is not readily available. It require suitable infrastructure to absorb and to develop technology and to manage it properly and systematically. So to choose and adapt biotechnology can serve as a measurement of sustainable self-reliance that will allow it to participate fully in worldwide efforts to achieve sustainable development.

Technologically advanced developing countries such as the people's republic of China, India, Republic of Korea and Singapore in Asia and Brazil and Cuba in most of these countries have invested significantly in infrastructure and human resource development and have increasingly encouraged foreign investment. The result has been the establishment of biotechnology based enterprises mainly fermentation industry and pharmaceutical products.

Biotechnological research programmes have steadily increased especially in agricultural sectors such as bio fertilizers, bio pesticides and virus free seedlings including various aspects of tissue culture. In developing countries biotechnology applications ranges from the use of advanced biotechnological techniques as *e.g.* in the production of transgenic crops and artificial seeds in China and in the production of several pharmaceutical products in Cuba and the Republic of Korea, to the use of traditional and intermediate biotechnologies in food fermentation and nitrogen fixation in less advanced countries. Several countries in the Middle East have emphasized the importance of biotechnology in developing stress tolerant agriculture and in bioremediation.

In African countries the level of complexity in biotechnology development is extremely unpredictable, ranging from very traditional application such as food fermentation of Cassara in least developed countries, to monoclonal antibody and diagnostics research and tissue cultures in other countries. Egypt, Kenya, Nigeria, Zimbabwe and the Republic of South Africa are among the leading countries in this area in the region biotechnology research and development in Africa evolves around the various international research and development centers as *e.g.* the International Institute of Tropical Agriculture in Nigeria and the International Center for insect Physiology and Ecology in Kenya.

Egypt and the Republic of South Africa have in particular given high priority to biotechnology in the countries development planning especially in relation to infra structural support and human resource development.

Economic Collision of Biotechnology:

Biotechnology lowers production costs, reduces dependence on energy from non-renewable resources, and increases the productivity of agriculture, lowers environmental and pollution problems.

Biotechnology differs from traditional forms of technology in three vital respects.

1. The creation of biotechnology involves input from basic scientists such as molecular and cell biologists and cytogenesis that are not port of traditional agricultural research extension network. These scientists are not in colleges of agriculture or agricultural research stations and they may have little, if any familiarity with agriculture. This poses a problem in establishing research priorities that reflect the needs of farmers and consumers.
2. The process and/ or products of biotechnology are often patentable. This stimulate private investment in agricultural research, but bulk of agricultural research was conducted in the public sector today private sector investment in biotechnology related research exceeds public sector investment.
3. Research on biotechnology is expensive simple forms of biotechnology research *e.g.* tissue culture costs US $ 1 million and takes 3-6 years. However, genetic engineering or gene transfer requires between US $ 50 million and one to four decades achieving the goal.

So basic biotechnology research is too costly and too demanding of scientific skills for the inadequate assets of most of countries

According to Banker R (Connell University USA) biotechnology will be of benefit of developed and developing countries comparable, the gap between favorable and unfavorable forms, regions and national likely to grow, for the following reasons.

1. The difficulty in transferring knowledge intensive technology to under developed areas.
2. Limited access because of the increasing privatization of research
3. Loss of export markets for developing country commodities through substitution in end products.

Today much of the developing world depends on public sector investment in agricultural research and extension, but budgets for research and the quality of national research institutions have declined in many developing countries over the past decade.

Support from developed countries and donor agencies are declining with the growing pressure for assistance to Eastern Europe. There has been a tendency to extend investments in research and extension over a large number of institutions; rather than developing a few quality institutions. It weakened capacity for technology transfer and hence reduced ability to adapt biotechnologies to local conditions

Biotechnology offers the prospective for further substitutions in end product, which could reduce the demand for export from developing countries.

The failure of the General Agreement Trade and Tariffs (GATT) negotiations to reach agreement on reducing agricultural protection in the developed countries creates on additional incentive to produce substitutes *e.g.* liquid maize sweetener has replaced about 50 percent of the sugar consumed in the USA and accounts for 15 percent of the world's sweetener market. The high price support for domestic sugar in the USA provided the incentive for developing liquid maize in demand for sugar cane exports from the developing countries.

Biotechnology focuses on the use of end products such as starches, proteins, cooking oils or sweeteners and thus makes it easier for scientists to respond to distorted prices by creating cheaper substitutes.

The African countries that reply principally on exporting tropical products and importing food are the most exposed.

Economic Expansion

In USA, many new biotechnology companies were founded, funded by undertaking capital. New biotechnology companies are defined as capitalist ventures, which are recognized during and after 1970 whose function is research development, and production using biotechnological means Genetech was the first biotechnology company, which started in 1976 in the USA.

In 1980 the US Supreme Court made the landmark ruling that micro-organisms could be patented thus on condition that the essential protection and encouragement for viable development of biotechnology products in 1981 Japan and France had followed the early biotechnology initiatives of the USA, UK and Germany and the first monoclonal antibody diagnostic trails were standard for accomplishment in the USA; where so biotechnology companies had previously been recognized.

In 1970, trans-national chemical companies which produces, chemicals, pharmaceuticals and farm chemicals fertilizers and pesticides made their investments in biotechnology there importance in biotechnology keen because of amplify in oil prices during the crisis in 1973, which established the risk associated with the heavy reliance on petrochemicals.

Mounting responsiveness of environmental and regulatory issues related to foam chemicals also provided a stimulus for industry to search for alternative strategies and products that would be more appreciate for the 1990s.

Supplementary transnational chemical companies came to similar conclusions and decided that significant investments in biotechnology represented substitute approach that would not be vulnerable to sudden changes in oil prices.

This alternative biotechnology strategy was adopted to generate high value drugs and agricultural products and would be appropriate from a regulatory/environmental viewpoint, and would contribute to significant economic increases in agricultural productivity.

In the late 1970s and early 1980s companies such as Monsanto and Dupont invested significant sums, of the order of US $150 million for biotechnology applications in human health, plant and animal sciences. This early trend in the USA was mirrored by similar developments in Europe based transnational and followed by Japan based transnational. During the 1980s the chemical transnational are also estimated to have spent an additional US $10 billion acquiring seed companies which are designed to provide the marketing distribution system for their plant biotechnology products.

Biotechnology Approach

The activity of private sector in biotechnology is becoming incorporated and consolidated on a global scale. This is because high R&D expenditures requires to be successful in this highly competitive field, thus integration achieved between private sector corporations but considerable progress is being made with forging links between the private and the public sector, a move that is being promoted or even required in many cases by government policy.

Industry views public sector research and the need to foster its continued growth and development as critical for the sustained growth and development as critical for the sustained growth of biotechnology. Public sector research of the kind that initially founded biotechnology; it support the calculated and applied research conducted by industry in the course of developing biotechnology products, for the market place research at universities and teaching institutions is also vital as it provides the necessary milieu for academic staff to work at the cutting edge of research and equally important to train tomorrow's biotechnologists. The private sector provides significant support for public institutions and universities to conduct research under contract in return for patent rights often with the provision that some portion of royalties may revert to the public sector laboratories.

Targeting Policies in Biotechnology

It is defined as any policy that singles out the indigenous development of biotechnology for special attention from the central government, identifies which targeting mechanisms could be adopted and institutions comprehensive programmes in order to help domestic companies to develop certain areas of biotechnology. The targeting policies reduce economic risk and lessons corporate duplication in biotechnology research and development. Some countries have designated the commercial development of biotechnology as essential to their nations continued economic well-being. Using a defined policy of funding research in the life sciences and encouraging research and development (R&D) in all industries with tax credits, these governments have instituted targeting policies in biotechnology designed to promote the commercial development of biotechnology.

Government Endowment for Fundamental and Functional Research

The intention of research is to gain a better understanding of the fundamental aspects of phenomenon without goals towards the development of specific products or processes. It is critical for maintaining the science base on which a technology rests and to stimulate advances in a technology. Academic researchers who receive government funds conduct this basic research. The objective of applied research to gain the knowledge needed to supply a recognized and specific need through a product or process; these are funded by industry.

Genetic research included development of bioreactors, screening of micro-organisms for potential products and better understanding of the genetics and biochemistry of industrially important micro-organism support of basic science and generic applied science is generally viewed as the responsibility of government because it ultimately contributes to the public good and because it is high risk and too expensive for individual firms.

There is controversy exist regarding relative importance of national support of basic and applied science. Some of national support of basic and applied science, some said that finding of basic research are readily accessible worldwide because they are published in journals with international distribution, strong government support for basic research is therefore not required for the maintenance of a leading portion in the development of a technology. Other said that the development of a technology within a country would progress faster if companies have access to local basic research scientists for consulting and contractual arrangements. Domestic technology transfer can help give industry a lead in innovation.

Individual Accessibility and Training

For the industrial development biotechnology the vital factors, which play important role, are trained scientific and technical personnel's. Countries which are not having highly skilled personnel's cannot compete internationally in highly technical operations such as the design and manufacture of a computer controlled bioreactor, the discovery of a new biochemical pathway for the production of specialty chemical or the development of a micro-organism that produces a desired protein. Another important factor for the success of companies commercially is the degree of sophistication of their research and development (R&D) personnel's with respect to state of the art developments in the field. There is no typical firm or organizational structure along the firms using biotechnology, most corporate activity in new biotechnology at present dedicated to R&D. as commercial development is totally dependent on skilled personnel's companies offering special inducement to highly qualified personnel's; as well as they also gave them considerable freedom with respect to the pace and direction of their work. Important categories need specialist such as in rDNA and MAb technology required molecular biologist and immunologist. The availability of the scientific and technical personnel's necessary for the commercialization of biotechnology is highly depend on a country's educational infrastructure at the following levels such as, secondary school education, biotechnology-related undergraduate and graduate education, translational training opportunities and mid career retraining opportunities.

University/Industry Association

Main purpose of university is education but with this conservation of knowledge and the per suit pf unrestricted knowledge is also important. In contrast to the function of universities, the object of industry is to make a profit and the mode of achieving this target is antagonism. Fundamentally industry is output oriented *i.e.* it aspires to be efficient production of goods and services.

When company pays for research it may expect ownership of the results long enough to justify the investment to bring the product to market.

In industrial setting there is less willingness than there is in a university setting to share research materials, which are often kept as trade secrets. The reason for secrecy in industry is that development of a product is risky, costly and fraught with many obstacles along the route of success. The cost of developing and marketing a product may vary among industries and products, the development of a pharmaceutical product in the US can cost from $50 million to $75 million with no guarantee of profit. In industry achieving competitive edge in a market necessitates guarding communication and intellectual property an operating mode is quite opposite from that of universities.

As industries and universities undertake partnership in biotechnology because of variety of reasons ranging from the desire by industry to gain access to new technology, to gain a lead time in basic knowledge, or to obtain trained personnel, to the need by universities to fill shortfalls in funding It is hoped that the effect of the partnership will be to facilitate and speed up the progress of domestic technology transfer, as it is critical to the maintenance of a competitive position.

Optimal technology transfer between universities and industry is a two way synergistic process between equal partners. Basic research is usually carried out in universities and further advances in technology is developed by industry *e.g.* recent research in molecular biology could not have been made without advances such as electron microscope, X-ray crystallography, etc. thus universities and industry alike must accept the requirements of the other institution and enter into agreements that maximize the ability of each to maintain its standards and goals.

University/ Industry interaction is a very effective way of transferring technology from a research laboratory to industry. These interactions promote communication between industrialists and academicians and it is also beneficial to both sides. Industrial scientist learns the latest techniques and research results, while academician gain increased familiarity with challenges to industrial R&D.

It also increases financial benefits to university research by industrial funding. There are disadvantages also, but can be avoided if both are aware of potential difficulties and adequate safeguards are in place one major disadvantage of closer relationships might be a tendency to increased secrecy on the part of university faculty; it should be noted that some secrecy always existed when a particular faculty member is close to a new discovery.

A second major disadvantage is the danger that basic research in universities will be directed towards profitable lines of inquiry instead of toward interesting questions raised by past or recent finding. This happens when there were a precipitous decline in national support for research in universities and universities had to turn increasingly to industry for financial support.

A third disadvantage is that some universities might be associated with products and processes linking between them to lawsuits for damages, causing subsequent imprisonment of the universities impartially and credibility. Finally there is the danger that university that traditionally has competed for the best faulty might compete instead for the most lucrative industrial contracts.

As biotechnology increased university/ industry interaction, many university/industry agreements in biotechnology focus on research directed toward applications of biotechnology and in a specific industrial sector, whereas other university/industry agreements are directed at many applications of biotechnology.

Fitness, Protection and Environmental Parameter

Instruction has been and will continue to be a factor in the development of biotechnology in particular for recombinant DNA or rDNA processes and products. When the rDNA technique was first developed its novelty tremendous power to manipulate organisms raised the specter of potentially drastic penalty to human health and the environment through the creation and proliferation of organisms with unknown but potentially hazardous traits.

As scientists learn more about molecular genetics risk associated with probing the unknown diminished and no evidence was discovered to support many of the early risk scenarios. Molecular biologist gained the confidence of the public by bringing other experts and the public into the decision making process that established the system of voluntary self-regulation. There has been no evidence of any harm to human health or the environment from rDNA. Questions have been raised about the long-term effect on worker's health from exposure to novel organisms and products and about the risks of deliberately releasing genetically manipulated organisms into the environment. There are other products also which has risk.

Special risk may lead to limited new regulation that could direct commercial efforts away from certain areas or at least slow advancements in those areas. Some pharmaceutical and chemical products that could be made with the help of biotechnology already subjected to considerable regulation. These regulations affect corporate strategies and patterns of industrial development. The cost and time involved in complying with regulatory requirements are the price society pays for society. Those countries, which have most favorable regulatory environment interns of least restrictions and uncertainties, will have a competitive advantage in commercialization of biotechnology.

Four Areas of Parameter are Considered

1. Parameter directed specifically towards biotechnology
2. Environmental parameter, which is relevant to biotechnology.
3. Existing parameter that would apply to biotechnology products.
4. Worker health and safety parameter

The strategy for rDNA research imitates the decision by experts and policy makers as it represent some special risks and uncertainties that require special attention.

These are based on, rDNA research should be conducted at increased levels of physical and biological containment related to the degree of possible hazard and the degree of oversight should relate to the degree of possible hazard.

Generally biotechnology products subject to existing regulation for generic products, it is necessary to assess these regulatory regimes.

Intellectual Property Law

Biotechnology will give rise to vast display of new inventions; these are placed into two categories, products and processes products include organisms, genetically modified micro organism, cell lines, hybidomas, plants, animals, parts of animals, plasmids, viral vectors, synthetic genes, probes and restriction enzymes. Processes include, various ways to make new organisms or parts, or use of organism to make product such as insulin.

Different bioprocess techniques such as plant tissue culture, breeding techniques The aptitude to secure a property interest in an invention and to protect related know how an extremely important incentive for a private company to spend time and money to carry out research, development and scale up for the commercialization of new processes and products. Without the ability to prevent other companies from taking the results of this effort, many new and risky projects that could lead to important new products would not be undertaken.

The ability to secure property interests in or otherwise protect technological processes, products, and know how will encourage development of technology

There are three categories, of intellectual law, which are related to biotechnology; dealing with trade secrets, patents, and plant breeders' right copyright protects the tangible expression of information and a gene, which may be viewed as the tangible expression of information. Other laws relevant to biotechnology are personal property law and contract law.

Traditional personal property law will apply to cell lines and many other biological inventions, because they are physical objects. Thus biotechnological inventions can be protected by contract and in view of some of the uncertainties in the intellectual property law regarding biotechnology, contracts can be important to biotechnology companies in may instances.

Bio Safety Strategy

Developing countries must have their own guidelines as well as global guidelines. It should be flexible and change with as new knowledge comes in. it is very important that bio safety regulations make certain that no scientist in his/her eagerness carries out research without suitable safeguards and containment facilities. Scientists themselves should evolve codes prescribing what should be avoided and what can be done. Some kind of internal code of ethics in biotechnological research is needed.

Patent Policies

Developing countries must develop their own patent policies. They should not be thinking of themselves all the time as receivers, but should give thought to their role as innovators and giver. Developing countries can, thus, operate patents in a variety of ways; a public sector unit may utilize a patent purely for public purposes. What is important is a clear policy on the role of patent in national and international technical co-operation.

1. Interaction between the public and private sectors, between the academia and industries, and between scientists and technologists.
2. Technologies must promote, not endanger, the livelihood security of landless labor families and those without assets.
3. A careful identification of priorities in R&D investment based on comparative advantages in absolutely essential because of resource constraints. The poorer the resources, the more important it is to have a correct sense of priorities.
4. Competitiveness both in terms of negotiations in GATT and in terms of efficiency and quality of products.

5. Substitutes for natural products may threaten agricultural income and trade in some cases. Technology alert systems should analyze the social and trade implications of such technologies.
6. Ecological security, that is, the protection of long term productivity of land, water, flora and fauna and atmosphere must be primary goal of all R&D efforts.

Biotechnology Revolution

Biotechnology revolution in the Third World can be done through four stages:

1. The transfer of technology by obtaining licenses for production.
2. The integration of existing technologies to design and produce new products not yet available but needed by the market.
3. The development of new technologies, when existing technologies proved to be insufficient for further development.
4. The undertaking of basic research to support the previous stages and to maintain technological abilities and competitiveness.

These four stages of biotechnology revolution will overlap during the process of development.

Action Plan

The development of biotechnology in the Third World, as part of the social and economic development, should establish the following.

a. The development of bio-industries for producing goods and services for human welfare to support national development.

b. The development of research centers able to carry out advanced biotechnology research and development to support the bio-industries in maintaining and improving their capability of producing goods to fulfill the needs of the competitive market.

c. The development of educational institutions to produce the necessary expertise at various levels in biotechnology in support of the development of bio industries and research and development institutions.

d. The development of a co-operative network for biotechnology institutions at national, regional and international levels.

The approach of development should be "to start with the end to end with the beginning" that is, to start with the down stream processes and to end at basic research up-stream. With this approach, efforts must be made to reach the targets, such as:

e. To mobilize all national capabilities for the development of biotechnology.

f. To mobilize, maintain and improve the existing bio industries.

g. To encourage the use of established biotechnology processes to develop bio-industries.

h. To encourage, enhance and develop capabilities in research and development for advanced biotechnology to support the bio-industries.

To guarantee the efficiency of the development of biotechnology, all resources should be directed towards selected priority areas. The selected priority areas are agricultural, medical and industrial biotechnology.

Because of the numerous possible application of biotechnology it is necessary to select priority areas using the following criteria.

1. The expected products, whether goods or services, should meet the demands of the domestic market with the prospect for export in the future.
2. The availability of resources such as,

a. Basic materials
b. Executing institutions, government of private and
c. Executors, which include existing experts or those still in training.

General Conclusions and Action Plan

Biotechnology is not a new field of science but a combination of new tools and technologies arising from continuous advances in the scientific understanding of biotechnogical systems and process. The hard core of biotechnology is molecular biology. Many of the new tools now available come from research in the fields on molecular biology, genetic engineering and cell and tissue culture.

Appraisal of the potentials and limitations of biotechnology is now becoming feasible. It is clear that the last eight years have witnessed impressive progress in various fields of cell manipulation, particularly of genetic material, and there is no doubt at all that the 1990s will be an even more dynamic decade. It is also clear that we are on the threshold of many new advances of considerable practical significance.

Biotechnology has wide-ranging implications and, in some cases, already demonstrated benefits and impact, for example, in animal husbandry, some areas of crop husbandry, in particular, tissue culture; in micro-propagation health, industry, energy and environment.

The role of biotechnology, the emerging technology of the future, cannot be separated from the role of science and technology in general in the national development programmes. Science and technology should not be seen as just another aspect of development but it should be recognized as the vital force in any social economic development. Science and technology serves as the driving force in further technological development and plays a leading role in sustaining the high rate of economic growth. The role of science and technology in developing countries is quite different from that prevailing in developed countries. This difference originates from the disadvantageous position of the developing countries in their dependence on the more advanced countries for technology, and from the various internal problems rooted in the farmer's specific economic and social conditions.

Biotechnology development in the third world should address the following issues;

1. A package of technology, which is most appropriate to the country, including the blending of traditional and new technologies.
2. Packages of services, which are needed for all people to benefit from the technology and for ensuring that the economic benefits of new technologies, reach the undetached.
3. Policies on personnel and infrastructure development, which can attract and retain outstanding scientists, and make possible to build institutions around them.

Public policies will have to pay attention to infrastructure development in terms of training, re-training, educational and communication strategies, and awareness-generation at the public, political and professional levels. Research strategists should indicate what can be realistic expectations and the nature of the institutional mechanisms and personnel policies needed for strategic and applied research, the critical mass of efforts and facilities required, and above all, the role of multi-disciplinary efforts, international and regional collaboration and industry0-university collaboration. Industries and universities should establish technology, development and dissemination centers jointly.

The whole area of information dissemination, education, and mass media support is critical to success. Mass media can help to generate the necessary political will and political action.

Chapter 2

APPRAISAL OF BIOTECHNOLOGY

The appraisal of Biotechnology is as an interactive commitment-building process. Technology appraisal is the magic infuses which enables government to guarantee the total quality and relevance of the technological developments they approve or stimulate. Other initiators of technologies such as biotech companies or academic research labs, make use of the same tool to convince the authorities of how well prepared, beneficial and safe their plans are technology appraisal also provide activities and environmental first with an infallible instruments to collect irrefutable evidence of the danger of specific developments which they had since long suspected. The second point is that the technology appraisal ideas are a permanent treat to any serious attempt to monitor the relevance and harmlessness of technological developments. By holding up the prospect of providing a complete and objective analysis. Technology appraisal plays along with our deeply human but utopian craving for making infallible decisions and in this way directs our attention from more realistic approaches to enhancing the quality of decision making under uncertainty.

This chapter begins with an explanation of the uncertain situation of the golden sixties during which the American Congress conceived the suggestion of Technology appraisal and coined the term.

Sixties were dominated by technology. First it was seen as the driving force behind wealth and abundance and later as a potential danger and a major source of an irrelevant and degenerating lifestyle. Presentation of Technology appraisal is the attempt by US congress to identify valuable technology and to separate the wheat from the chaff. This outline is followed by an evaluation of this enterprise explaining the amenability of the approach and paying attention to the positive lessons we can learn from the Technology appraisal experiment.

Reimbursement of Technology

New technological developments in the 1960s endure the idea that are in fact suspicious and in need of critical assessment before they be allowed to proceed. Technological innovation was believed to be a good thing, which should be stimulated by the means. Important aspects of social life appeared to have revitalized all by Technology

Growing Social and Economic Welfare

Never before sixties the people in Western industrialized countries experienced a similar wealth and comfort since World War II, the average productivity of human labor, the Gross National product and the GNP per capita had steadily been increasing by about 3 to 4% per year. This trend of growing prosperity not only resulted in more dispensable income for the families' money to invest for national states but also is a decrease in the amount of working hours, which in turn, allowed for more time for leisure.

Technological innovation, which is based on strong scientific research, manifested itself at the heart of new social order. It appeared that four hundred years of modern science and two hundred years of modern technology had finally reached their stage of maturity. As the beginning of the exploration and exploitation of promising fields and powerful new technologies much faith was placed in the fact that this process of increasing prosperity had been set into motion for good. Four fields were expected to lead to new revolutions with breath taking new possibilities.

1. The electronic revolution with the integrated circuit, which at the time was still in its infancy. Its application had already drastically highlighted economic productivity despite its newness. By the years 2000, it was expected that the computer would enhance the productive capacity of human labor by another 300%, allowing every person to do the work and to earn the income of three.
2. The development of new and cheaper source of energy. Promising new technology would allow the unleashing of enormous amounts of energy stored in the inner core of highly energetic material by converting this hidden energy into electricity for everyday use. This movement is also called the nuclear revolution.
3. Certain traditional basic material, like steel, as its elemental properties were difficult to engineer, could be replaced by more flexible man made materials such as plastics the properties of plastics present the advantage that they can be designed in a research facility and specifically adapted to their planned use, and at a later stage, they can easily be manufactured in chemical plants by employing simple readily available raw materials.
4. Biological revolution; genetics was the driving force behind this revolution. This discipline was seen as the study of the core encoded the characteristics of the different species and forms of life and which are used by living organism to transmit hereditary traits to later generations Research increasingly succeeded in unraveling the basic mechanisms of those mysterious units and subsequently opened up prospects for powerful technological applications not only in the field of human health care but also in agriculture and other fields of human activities. By the end of the sixties, science had already succeeded in isolating the first gene, but it would take until the seventies before

the first technological uses of those genes would become possible.

These prospects combined were the driving forces behind academic science and technology and growth of industrial companies. Only large companies could afford to be involved in the basic research which was needed to guarantee ongoing economic growth and it were only these same large companies that could afford to develop the mechanisms can to turn these basic scientific discoveries and technological developments into real industrial products to be distributed in society.

This beneficial effect of technology on industry makes life easy for mankind and gave happiness to individuals and families. New machines helped to reduce burdensome work and to organize and enrich daily life washing machines, dish-washers refrigerators and freezers became available and accessible to each household color television opened a more vivid window on the real world and provided a private entertainment center for every living room. Communication was facilitated by a growing number of telephones and cars. Because of this the golden sixties were filled with music, wealth and hope.

Maintaining Peace

New developments in technology left its mark on the widespread belief that there would never be another war like World War II. It was clear that war had not disappeared from the earth but at the beginning of the sixties massive war with its atrocities and misery did disappear from the daily life of westerners war. There was mere diplomatic tension between two blocks the cold war between the East and the West or to the theoretical knowledge of smaller, local conflicts in hidden parts of the world. There was a general faith that Western diplomats, politicians and military forces could master all existing threats because they could rely on the most sophisticated arsenal of high-tech weapons then known to man which would easily be able to stop all attacks from the outside. It was clear that these weapons were so powerful that no other power would ever dare to try to attack the West, because they realized that the West could immediately reply to any attack with the utter eradication of the attacking force and indeed its entire nation by means of a series of atomic bombs many times more powerful than the bomb which destroyed Hiroshima. It was believed that technology would be protecting the modern Western World. Refined weapons replaced bow and arrow while radars and satellites replaced the Watcher's eye.

Hidden Costs

Some technologies, developed and applied with only the best of intentions, threatened our natural and cultural resources long taken for granted.

Cost to the Natural Environment

In 1962 Rachel Carson published "Silent Spring". In this dramatic and alarming report, she convincingly demonstrates how the increased one of pesticides is slowly but surely poising our biological environment. It is the first alarming signal given to society that by undermining the environment our technological civilization could ultimately turn against humanity instead of promoting it. Her claim was followed by many negative reports about the effects of motorways, jet aircrafts and airports nuclear energy and many other phenomenon of our modern culture.Until that time it had scarcely been thinkable that the increasing industrialization would ultimately lead to the widespread extinction of aquatic life, to heavily polluted coast and beaches or indeed to sickening and unhealthy air pollution in the large cities. In the seventies when biotechnological methodologies were revolutionized, the scientists and technologists behind these break through had already expressed their deep concern for the potential hazards of the use and laboratory study of recombinant DNA molecule hazards to the human health as well as to the health of our environment leading molecular biologists even proclaimed a moratorium for all further research. This gesture of caution attracted the attention of the press and policy makers and fueled the widespread belief that the potential dangers associated with biotechnology are of a very special nature.

Cost to the Social Environment

A great diversity of threats to the Social environment was also increasingly considered unemployment was a manifest example of these social dangers. The ever-changing technologies in the workplace adversely influence the motivation of workers, as it could be feared that they would not want to consistently adapt to new procedures or to be in need of permanent education.

The technological implications of this new age the social environment in the work place, the cultural and family centered social environments were not spared. For example it had been argued that television had signified the death of social and cultural life and even of family life. Also the enormously increased mobility of the jet-age generation was described as a potential threat to family life. Our legal and social mechanisms to the many new and powerful technologies, such as the broadened potential of medicines in the fields of prenatal life and care of the elderly not everyone was convinced that the new social optimality in a better and more humane way than the old order did. Again the oncoming biotechnology and its multiple uses in agriculture, industry and health care stirred.

Rise of Technology Appraisal

The 1960s ended with a deep awareness of crisis. Technological and industrial development had proved to be accompanied by a host of social and environmental costs to which it had been blind for many years. National government which was they systematically involved in large scale technological development programs, felt compelled to seriously examine and reconsider their earlier carefree approach and their priorities.

The American Congress launched its own initiative, prepared by the subcommittee for science, Research and Development of the Science and Astronautics Committee of the House of Representatives.

Regulation of products and process of the new biotechnology by the United States National Institutes of Health (NIH) has been much to praise and also much to criticize on the positive side, oversight has been relatively enlighten with progressively decreasing stringency of their guidelines for laboratory research and the de facto elimination of NIH approvals of field trials. The approach of the United States Food and Drug Administration (CDA) to the new products, overseeing them in the same way as similar products made in other ways, has been validated for a large number of products with more than 800 (mostly diagnostics, plus approx 20 therapeutics) approved for marketing and more than 1500 clinical trials of drugs and biologics under way on the negative side, good science and good sense have not always been well represented in the formulation of regulatory policy and too often the bureaucracy builders have instituted burden some after its time. Several European countries and Japan have adopted a process-based approach to oversight and have been unnecessarily restrictive and they have paid the price in diminished research and development activity. In certain areas in the United States in hospital regulatory climate and a lack of public understanding have discouraged research and development. In particular, academic and industrial researchers have been leery of field trials of recombinant organisms' bio control agents that were once highly touted and investor interest in sectors that require field trials has diminished.

Majority of application of the new biotechnology are fundamentally similar, on both theoretical and empirical grounds to well accepted and the applications of the old biotechnology are demonstrably safe. But there are some new biotechnology products are hazardous or even potentially hazardous is unjustified and lead to a number of undesirable consequences. Therefore it is very difficult to acess these new technologies. These outcomes will in turn cause the public ultimately to suffer from a failure or delay in the development of new products.

New biotechnology is whether the trigger for regulatory jurisdiction or for a certain degree of regulatory scrutiny should be the use of certain process of genetic manipulation or whether the characteristics of the product (either a living organism or inert product) or its intended use should contribute the trigger. This is the process versus product debate, which has dragged on long after the issue should have been put to rest by international scientific consensus.

However not only most products are safe but also the public must have confidence in their safety. At the same time public misapprehensions cannot be permitted to stimulate regressive policies, thus the advent of the new biotechnology continues to pose a challenge that involves perception as much as reality and public education as much as the presence of effective and coherent regulatory framework.

There are some traditional beliefs about biotechnology.

Traditional Beliefs About Biotechnology

1. Biotechnology is a Discrete New Technology

Biotechnology is merely a catch all term for a broad group of useful, enabling technologies with wide and diverse applications in industry and commerce. This spectrum of tools has become so integrated into the work of practitioners such as plant breeders and microbiologists that is difficult and also pointless to attempt to distinguish between old and new.

A useful working definition of biotechnology is the application of biological systems and organisms to technical and industrial processes. This definition encompasses processes as different as fish farming, foresting, the development of disease resistant crop plant, the production of enzymes for laundry detergents and the genetic engineering of bacteria to clean up oil spills, kill insect larvae, or produce insulin.

Processes in Biotechnology are dissimilar products for vastly dissimilar applications. These and products are so diverse and have so little in common with one another that it is difficult to construct valid generalization about them for whatever purpose because of biotechnology's lack of systematic, uniform characteristics, it cannot be effectively legislated or overseen in the uniform all inclusive manner that is possible *e.g.* for the underground coal mining industry.

As the uses of Biotechnology products vary, so do agencies jurisdiction and the nature of their evaluation of the products .e .g. A review by the US Environmental Protection Agency (EPA) of an enzyme used as a drain cleaner will be different from review by the FDA of the same enzyme injected into patients to dissolve blood clots. The diversity of products and their applications argues against the usefulness of legislation or regulations that attempt to encompass unnatural groupings that would occur under general terms such as biotechnology genetically manipulated organisms or planned introductions.

2. Biotechnology and Genetic Engineering are New

Biotechnology is new. But on the contrary many forms of biotechnology have been widely used for millennia. Earlier than 6000 B.C. the Sumerians and the Babylonians exploited the ability of yeast to make alcohol by brewing bear. A picture of the ancients preparing and fermenting grain and storing the brew has actually been retained for posterity in a pictogram. Genetic engineering is an older subset of biotechnology the indirect manipulation of an organism's gene by the guided mating of animals or crop plants so as to enhance desired characteristics. In this process of domestication changes one can made at the level of the whole organisms *i.e.* selection is made for desired phenotypes and the genetic changes often poorly characterized, occur along with. The new technologies in contrast enable genetic material to be modified at the cellular and molecular levels. These new techniques yield more precise and deliberate variants of genetic engineering and they consequently produce better-characterized and more predictable results, however they retain the aims of classical domestication.

To yield many valuable organisms domestication of microorganism for the modification of foods, industry has used classical genetic methods. One example is the genetic improvement of Penicillium chrysogenum, the mold that produces penicillin; by several methods including the screening of thousands of isolates and the use of mutagens to accelerate variation, penicillin yield have been increased more than a hundred fold during the past several decades. Variety of other important substances; including industrial detergents, antibiotics, organic solvents, vitamins, a; poly-saccharides, steroids and vaccines can be produce by, microbial fermentation which is employed throughout the world.

The value of these and similar products of conventional biotechnology is in excess of $100 billions annually. With the identification of the genes responsible for the biosynthesis of various antibiotics the empirical search for improved products or yields is being supplanted by the introduction of specific directed genetic changes.

By adding up to such contained industrial applications there have been many beneficial planned introductions of other organisms into the environment, including insects, bacteria and viruses. Insect release was used successfully to control trouble some weeds in Hawaii in the early 20th century. Other examples are the successful program for the biological control of St. Johnswort (Klomath Weed) in California by insects in the 1940s and 1950 the more recent use of an introduced rust pathogen to control rush skeleton weed in Austria. There have been hundreds of planned introductions of natural predators of insect pests since the first such release of the Austrian Vedalia beetle in California in 1988. This area of research the biological control of weeds, nematodes, insects and diseases is actively promoted by the US Depts. of Interior and Agriculture.

At present more than a dozen microbial bio control agents are approved and registered by means of the EPA, and these organisms are marketed in hundreds of different products for use in agriculture, forestry and home gardening. In other major area-growth promoters for plants-preparations containing the bacterium Rhizobium, which enhances the growth of leguminous plants such as soybeans, alfalfa and beam, have been sold in this country since the late nineteenth century. These products provoke the development on plant roots of structures in which bacteria can convert N2 into compounds that can be used by plants and thereby decrease the need for chemical fertilizers.

Some of the most omnipresent intended introductions of genetically modified organisms have involved the vaccination of human and animals' populations with live, attenuated viruses. Live viruses modified by a variety of techniques and approved in the US encompass vaccines against mumps, measles, rubella, poliomyelitis, and yellow fever infection of the recipient but every now and then the possible of further transmission to the virus in the community with the acknowledged risk of serial spread. Yet no vaccine virus has become recognized in the environment regardless of the repeated occurrence of vaccine viruses there, *e.g.* studies in the US and the United Kingdom have shown that vaccine strains of polio virus are there in sewage but this reflects the abiding administration and excretion of live virus vaccine to a certain extent than its sequential propagation through human or animal hosts in the community.

Viral vaccines created with grown-up genetic techniques for attenuating virulence have been surprisingly effective all over the world and have fully eradicated the dismay smallpox virus. Regardless of rate unfavorable reactions the favorable blow of vaccines has been to this position old-fashioned in significance as a supporter of human legerity and value of life only by the agricultural green revolution, which was mediated in part by genetically, modified plant.

Building on the successes of the ancient times the most recent biotechnological techniques including DNA are already given that still more specific, better-understood and more conventional methods for manipulating the genetic material of viruses and other microorganisms for employ as vaccines. The modification of crop plants has been performed eternally in view of the fact that ancient agriculturists preferred for cultivation plants possessing desirable traits from domesticated links of wild species. The rediscovery in 1900 of Mendel's perception of inheritance to no purpose in the scientific relevance of genetic principles to crop improvement in view of the fact that then every scientific move on has improved the capability to improve as expected the genotype (and more important for the consumer of the product the phenotype) and now a grouping of several techniques is frequently used to improve plants. *E.g.* an existing plant might have been modified by many generations of classical breeding and selection and more in recent times by monoclonal variation and wide crosses with embryo liberate. By the newer molecular techniques these plants are being further improved and can then be reintroduced into classical breeding programs from its descendants will be released into commerce. By increasingly used inter specific hybridization in this century plant breeders have to transfer genes from convinced non-cultivated plant species to a range of different species. These inter specific transfers of traits from wild species to domesticated links into the same genus motivated attempts at even wider crosses, together with those between members of different genera. These wide crosses transcending usual barriers to mating have been facilitated by embryo liberate culture techniques in which a sexual cross compliant a useful embryo but abnormal endosperm is rescued by culturing the embryo. This is skillful by given that the hybrid embryo with the life support normally supplied in the early stages of development by maternal tissue and the endosperm. In different places a number of plants resulting from wide crosses have been used in additional breeding intensively field-tested and marketed. These plants include usually available varieties of tomatoes, potatoes, corn, and oats, sugar beets, rice and bread and durum wheat. Between unrelated organisms the molecular techniques for the genetic manipulation of plants facilitate scientist to move precise and functional segments of genetic material. These techniques offer several compensations and harmonize existing breeding efforts by increasing the diversity of genes and germ plasma available for incorporation into crops.

The frequent molecular techniques for genetic manipulation of plants can be divided into two types-vectored and non-vectored. The entry of the foreign gene into the plant cell by vectored modifications respond on the use of biologically active agents, such as plasmids and viruses to make easy. Non-vectored modification responds on the foreign gene being physically inserted into the plant cell by such methods as electroporation, microinjection or particle guns. Even though in some respects the use of these molecular techniques is more complex than traditional plant breeding practices it is likely to be at least as safe. In both kinds of approaches new DNA enters the plant's genome and is stably maintained and expressed.

3. Unknown Compensate the Known

As far as the ecological properties of microbes are concerned, the unknowns for prevail over the known it is exceptionally negative and even ambiguous to disagree. In detail it is not what is unknown per se that is material but what is unknown that relates to risk that is important. In the extract above, one could just as without difficulty replace with, the functions of mutations in live poliovirus vaccines for the ecological properties of microbes but for four decades these uncharacterized genetic changes have not prohibited our using empirically developed, live, attenuated poliovirus vaccines. The statement is particularly dubious for many micro-organisms of commercial interest, including several pseudomonas species (Syringae, aureofaciens, fluorenscens). The microbes used in fermentation to produce bread, cheese, wine, beer and yoghurt Thiobacillus species, Bacillus thuringiensis, B subtilis, Rhizobium and baculoviruses. In reality the mainstream of microbes are necessary to ecosystem processes or otherwise helpful to man and only a microscopic fraction are pathogenic or otherwise harmful.

In anticipation of the recent-furor over engineered (meaning rDNA manipulated) organisms, small scale field trials of native or non pathogenic microbes modified by classical techniques were long exempt from both the safety record extending over more than half a century is admirable. The scientific method and prior experience when applied reasonably to risk evaluation do facilitate us to make functional predictions.

4. Original and Hazardous Organism will be Created

The new genetic engineering techniques have been widely overstated by the degree of originality of microorganisms or macro organisms formed. A corn plant that has had newly incorporated into its genome the gene that synthesizes a Bacillus thuringiensis toxin is still after all a corn plant. E. Coli K-12 that has been programmed to synthesize human alpha interferon by means of rDNA techniques differs very little from unmanipulated siblings that manufacture only bacterial molecules. Furthermore in nature numerous recombinations via numerous mechanisms between even very indistinguishably related organisms have probably occurred. As in the gut, in decomposing corpses, and in -infected wounds bacteria can take up naked DNA, though inadequately and in nature they have long been exposed to DNA from disintegrating mammalians cells. The human population also excretes on the order of 1022 bacteria per day thus over the past 106 year's numerous mammalian-bacterial hybrids are likely to have appeared and been tested and redundant by natural selection. Even a single point mutation can modify virulence despite the fact that both genetic and ecological constraints function to prevent the emergence of remarkably pathogenic viral variants, (fields and Spriggs 1982). Whereas the variations that constantly arise on massive scale in nature do infrequently produce a modified pathogen, such as an influenza virus with increased virulence or the AIDS virus. With the tremendous background noise of recombination and selection in nature the chances of such an event arising from the small-scale changes made by man do not compare.

Because of the phylogenetic distance between donor and recipient have now been shown to arise in the laboratory and may occur in nature as well it is relevant that certain kinds of gene transfers thought until in recent times to be impossible in nature. *e.g.* Brisson-Noel *et al.* (1988) have confirmed that a gene or genes for erythromycin resistance was transferred between the Gram negative bacterium Campylobacter and not linked Gram positive bacteria. It was confirmed that gene transfer can occur between *E. Coli* and Streptomyces or yeast and that the crown gall disease in plants results from a natural transfer of DNA from a bacterial species (Agrobacterium) to plant cells in latest laboratory experiments (Heinemann and Spragne 1989, Mazodier *et al.*, 1989) (Information of the nature of this mechanism has already led to growth of a highly effective means of developing transgenic plants without danger of disease inductions, using DNA fragments from Agrobacterium). The difference between natural and unnatural is not a clear one and is arguably, irrelevant since evolution continually creates originality. Original neither is not one and the same with dangerous nor need it even mean apprehensively.

5. Nonpathogens will be Altered into Pathogens

Genetic manipulation unconsciously convert a nonpathogenic into a pathogen this is commonly spoken concerned about it. Still this view ignores the complexity and the multifactorial nature of pathogenicity. Pathogenicity need the evolution of a special set of properties that involve a number of genes whose functions control factors such as fitness, virulence and adhesion it is not trait created by some single omnipotent gene.

Three universal characteristics a pathogen must acquire are multifactorial

i) It must be able to metabolize and multiply in or upon host tissues *i.e.* the O_2 tension and pH must be adequate, the temp (and for plant pathogens, the tissue water potential) appropriate and the nutritional environment complimentary.

ii) The pathogen must be able to resist or avoid the hosts defense mechanism for a period of time sufficient to reach the numbers required actually to produce disease assuming that acceptance range for all of the many conditions necessary for metabolism and multiplications

iii) A successful pathogen must be able to survive and be disseminated to new host organisms.

The organism must be thoroughly modified to its pathogenic lifestyle. Unless many other required traits are present, even a gene coding for a potent toxin will not convert a harmless bacterium into an effective source of epidemic or even localized disease.

For human and animal pathogens these comprise at the least resistance to enzymes, anti-bodies and phagocytic cells in the host, the capability to hold to specific surfaces and the capability to increase on available nutrients provided by the host. No one of these traits confers pathogenicity, though a mutation that affects any essential one can eliminate pathogenicity.

In nature there are mild degrees of pathogenicity as severe pathogenicity is more demanding of favorable conditions and therefore much rarer. Consequently the probability of unintentionally creating an organism capable of producing a medical or agricultural disaster must be ravishingly small. The record of the laboratory use of rDNA techniques supports this prediction. For more than 15 years not a single adverse reaction has been observed in humans, animals, or the environment despite the release of the order of 107 recombinant micro-organisms per investigator per year from the std BL1 bio safety level laboratory (according to a 1983 analysis by Lincolin *et al.*, funded by the US environmental Protection Agency under Grant No. R-808317-01)

6. All Technology is Essentially Hazardous

The use of all new technology is hazardous as it disturbs the natural order and of breaking primitive taboos, combined with unfamiliarity with the statistical aspects of risk. The hazards of toxic chemical waste dumps and the technical problems of the nuclear industry (which are of uncertain significance to biotechnology in any case.) but they suitably pay no attention to the over whelming successes of telephonic communication vaccination blood transfusions, microchip circulatory and the improved productivity of plants, animals and microbes shaped by genetic improvement.

Processes or products of biotechnology are potentially hazardous in some way no one can deny this. Some of these hazards are already well known: workers purifying antibiotics have experienced allergic reactions laboratory workers have been infected inadvertently by pathogenic bacteria and vaccines have occasionally elicited serious adverse reactions.

Introductions of exotic species, such as English sparrows gypsy moths and kudzu have had serious economic consequences in addition. Organisms adapted by the new techniques with correspondence are made between the introduction of these nonnative (alien or exotic) organism and the possible consequences of field trials. On the other hand the evaluation is baseless, relying on the postulation that rDNA manipulation can modify the properties of an organism in a wholly changeable way that will cause it is affect the environment unfavorably. This conclusion makes very doubtful by both theory and experience. Usually introduction will be of native organisms changed minimally often by only the insertion or deletion of a single structural gene from organism that are already present in the environment and that will not take benefit from a selective improvement over their wild type associates. As their unmodified parents they will be subject to the same physical and biological limitations of their environments. What is more in the past, unusual nuisance organism whether introduced deliberately or unconsciously arrived in a new environment ecologically fit or modified and with their objectionable properties present. In disparity there is no indication that any organism has been made a nuisance by breeder's genetic manipulation of plants, animals or microorganisms. Consequently for organisms adapted by the techniques of the new biotechnology; the grown-up system of selective breeding, testing and case of domesticated plants, animals and microbes, with its long history and estimable safety record, is a more appropriate model than the introduction of exotic organism into environments in which a constructive niche existed.

Multifaceted and wide-ranging dictatorial apparatuses based in frequent governmental agencies in the US, European countries, Japan and elsewhere have long overseen the safety of food plants and animals pharmaceuticals, pesticides and other products of new biotechnology these authoritarian mechanism will persist to be equal to the task and to perform in a way that does not strangle innovation.

Chapter 3

OVERVIEW OF BIOTECHNOLOGY INDUSTRY

In the 21st century, biotechnology performed a revolution in many research-based sectors, carrying out a new approach to biological and chemical sciences.

It is recognized that modern biotechnology has been started in 1953, when nature published James Watson and Francis Crick's double helical structure of DNA, the chemical base for genes proteins, and cellular processes.

So modern biotechnology can be defined as collection of technologies *e.g.* genomics, proteomics, computational biology and chemistry and high through put screening which causes a rapid advance in all the traditional life sciences, particularly in the pharmaceutical R&D processes.

After Watson and Crick's work it took more than twenty years for establishing the first biotech company namely Genetech in San Francisco bay area. Four years later first product has been introduced in the market- genetically engineered human insulin.

The US legal context, which has been followed by other countries, evolved to adapt itself to the biotech scientific revolution.

In 1980 the US Supreme Court granted the patent ability of genetically engineered life forms and the congress approved the Bayh dole act, allowing scientists from different universities to fully exploit intellectual property rights on their research.

Applications of Biotechnology in Different Sector

1. In pharmaceutical sector, biotechnology products gives quicker and more accurate diagnostics test, more effective therapies with fewer side effects as they are based on the body's self healing capabilities with new and safer vaccines.
2. In animal healthcare, biotechnology is beneficial
   a. Improves animal health products.
   b. Advances in human health through studies on animals.
   c. Enhancement to animal products *e.g.* modified proteins or enzymes.
3. In agriculture industry, biotechnology can help to meet the ever-increasing need by increasing yields, decreasing crops inputs such as water and fertilizers and providing pest control methods that are more compatible with the environment.
4. In food industry, biotechnology provides new products, lowering costs and improving the microbial processes. It improves the quality, nutritional value and safety of the crops plants and animal products in addition to this biotechnology offers many ways to improve the processing of raw materials into final products, natural flavors and colors, new production aids, such as enzymes, emulsifiers. Improved starter cultures, more waste treatment options, greener manufacturing processes, and more options for assessing food, safety during the process.
5. In the manufacturing industry biotechnology employs the techniques of modern molecular biology to reduce the environmental impact of industrial process.

Industrial biotechnology makes manufacturing processes more efficient for industries such as textiles, paper and pulp and specialty chemicals. Some observers predict biotechnology will transform the industrial manufacturing sector as much as it has changed the pharmaceutical, agricultural and food sectors. It is a key to achieving industrial and environmental sustainability. There are many possible applications of biotechnology but currently the pharmaceutical sector is the most affected by biotechnology it redrew the R&D and even production processes allowing scientists to understand and actually exploit the genetic pathways to treat pathologies. It reshaped the pharmaceutical industry's structure.

Biotechnology and Pharmaceuticals

The biotechnology and pharmaceutical industries are among the fastest growing and most research and development concentrated. Changes in demographic uniqueness have increased the demand for products that remedy or assuage medical problems associated with aging, while rapid advances in knowledge of underlying biological and chemical processes have resulted in a revolution in the development of medical diagnostics and treatments.

There is noteworthy growth in service in its smaller bio/pharma firms in the areas of pharmaceutical preparations (45%) and commercial research (111%) from 1994 to 1997. This region shows a strong concentration of companies in the bio/pharma industry, with more than 50% more employees than would be predicted from its population. Significant growth was seen among smaller companies in the region among noncommercial research firms. There are more than 1,000 employed in the bio/pharma sector. Significant growth was seen among smaller companies in the region among manufacturers of pharmaceutical preparations and laboratory analytical instruments. Major bio/pharma companies, including Bausch and Lomb, Genencor, and Medeva Pharmaceuticals have operations in the region. The University of Rochester ranks among the top fifty Universities in the country in biological sciences based on the National Science Foundation's rankings of university research and development expenditures for 1997.

In General there are

- 34% of all Biotechnology firms
- 38% of all Medical devices firms
- 35% of all Pharmaceutical technology firms
- 33% of all Life systems technology firms

A number of developing countries now have well-developed biotechnology programs; they are imminent the most important periphery of biotechnology applications and have significant research capacity, according to a new FAO assessment on the status of research and appliance of crop biotechnologies in developing countries.

Based on a review of the information in the FAO database on Biotechnology in Developing Countries (FAO-BioDeC), which covers both genetically modified (GM) crops and non-GM biotechnologies, the assessment suggests that developing countries will soon have new GM crops available such as virus-resistant papaya, sweet potato and cassava, as well as rice tolerant to abiotic stresses (salinity and drought).

Most of the GMOs commercialized so far in developing countries have been acquired from developed countries and focus on a limited number of traits (mainly herbicide tolerance and insect pest resistance) and crops (commodities such as cotton, soybean, and maize).

However, the FAO assessment reveals that several developing countries have been conducting research on a wider range of crops, such as banana, cassava, cowpea, plantain, rice and sorghum, and on traits relevant for food security, such as a biotic stress acceptance and eminence.

Argentina, Brazil, China, Cuba, Egypt, India, Mexico, and South Africa have taken the front. A second group of countries has medium-scale agricultural biotechnology programs, usually in a few key areas. Other developing nations have relatively limited research capacity, according to the FAO report.

There are, on the other hand, some perceptible gaps in research. For example, no research is reported in the field of nematode resistance regardless of the considerable losses caused by these plant parasites. Another fundamental but abandoned research problem concerns post-harvest victims.

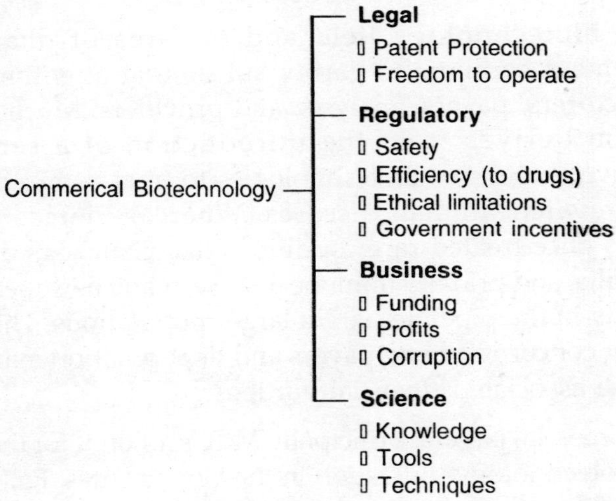

Fig. 1: Factors Affecting the Biotechnology Industry

The study also notes that bio safety competence building is needed to enable many countries in Africa, Eastern Europe, Latin America, and the Near East to benefit fully from GMO technology.

Regarding non-GM biotechnologies, many are being used on a commercial scale but only a few studies have been carried out to assess their socio-economic impacts. The report highlights that this is an area needing urgent attention, as it is likely to help guide research and technology policies and investments towards wider and efficient utilization of all biotechnologies.

FAO-BioDeC Launched in 2003 as an online searchable database, FAO-BioDeC currently has about 2,000 entries from 71 developing countries, including countries with economies in transition.It is regularly updated and has recently been expanded to include extensive data from the forestry sector and some initial data on livestock.

Five Steps of Biotechnology Management

The biotechnology field and the area of pharmacy management continue to garner substantial attention from manufacturers, payers, analysts, and providers. Much of this attention derives from the introduction of a series of innovative, expensive biotechnologies to treat comparatively more prevalent human diseases. Whereas biotechnology products once treated rare conditions like Gaucher's disease, hemophilia, and primary immune disease, many new therapies- and many in the pipeline- target larger populations. This shift is raising concerns among payers and depopulation managers about the associated financial burdens.

Commercial payers- principally MCOs account for the most of the biotechnology utilization in the United Sates. Following are the some stages discussed.

Each of these stages introduces new tools and strategies for reducing total biotechnology costs. As an organization moves from stage to stage, the complexity of the strategies grows, necessitating better management tools, diagnostics, and data to achieve desired objectives. More importantly, each stage increases the pressure on manufacturers to demonstrate the value of their products- or to make price concessions. Our research suggests that in several categories- particularly multiple sclerosis, rheumatoid arthritis, and hepatitis C - payers have reached the end of second stage. Future saving will necessitate more complex strategies and will present manufacturers with some challenging question.

(a) Procure and Schedule

Until only recently, most payers relied on their network or affiliated providers (physicians or specialty Pharmacy vendors) to obtain specialty products, manage inventory administer the product, and submit a claim, in return for managing the process, providers were able to bill at a premium above average wholesale price (AWP), ensuring a substantial renew stream. Some physician practices, particularly oncologists, relied on drugs to generate more than 65 percent of total practice revenue. The buy - and bill model provided payers with little control of costs or utilization.

(b) Area of Expertise in Pharmacy Self-administration and Compensation Changes

Faced with much greater costs compared to those associated with small molecule products, payers have taken some preliminary steps to reduce their exposure to biotechnology costs. By signing contracts with specialty pharmacy providers (SPPs), payers have been able to eliminate physician drug mark-up and, in some cases, have negotiated favorable pricing. The emphasis on self - administered therapies similarly allows payers to eliminate payments to physicians for office visits and administration while fostering patient convenience. This preference for self - administered therapies places greater pressure on Manufactures of infused agents to justify their comparatively greater cost. The penetration of SPPs into the commercial payer market - currently, 78 percent of payers use a specialty provider for all or some of their specialty products- has helped to drive acquisition costs down. At the same time, the passage of the Medicare Modernization Act (MMA) provides commercial payers with the cover needed to reduce payments to physicians still further. Our research shows that by mid 2004, the national buy-and-bill payment rate had quickly fallen to AWP minus 14.21 percent, effectively equal to the Medicare rate. AS Medicare com plates the transition from AWP based reimbursement to average sales price (ASP) this year, most payers expect to follow Medicare's lead.

(c) Grouping Contraction and Chosen Products

Opportunities to extract additional savings from providers have entered the area of diminishing returns, particularly as the MMA erodes practice profitability. As a consequence, payers will need to look elsewhere for savings opportunities, primarily by applying the lessons learned from small molecules to biotechnology and other specialty products. Comparatively few payers have meaningful policies in place that are advantageous to a particular biotechnology agent or that narrow a category to only one or two effectively comparable therapies. Exceptions include Aetna, Cigna, and Keystone Mercy, which have moved to narrow their multiple sclerosis product choices, and well point, which recently narrowed the selection of TNF-a inhibitors on its formulary from three to two. In other cases- human growth hormone and intravenous immunoglobulin. To date, most payers have not successfully moved patients to particular product, and the differences in patient cost sharing between preferred and non-preferred therapies have been minor or nonexistent.

Payers are exerting more pressure on manufacturers. Nearly 80 percent of payers agree that they would trade preferential position for a volume discount and, in many cases, non-preferred agents may not have access at all. Thus, if questions remain about payers' ability to move share under a contract, few doubt their ability to block reimbursement or coverage for individual therapies. For biotech products whose annual costs exceed $10,000, payers believe that narrowed categories are an attractive strategy to reduce biologic spending further.

(d) Advantage Intend and Cost Distribution Innovation

National data show clear evidence of rising patient cost-sharing requirements, irrespective of how therapies are classified, close to half of commercial payers now use tiered formularies to manage some or all biologics, generally placing these therapies in the highest tiers with co-payments that average nearly $40. Flat or no tiered co-payments now average more than $47, and cost-sharing requirements will continue to rise at a rate of 10-15 percent annually. As one HMO medical director told us, the "bottom line is: if costs continue to rise, co-payments will continue to increase, in disorders where it's not life threatening, such as rheumatoid arthritis, cost-sharing will get to the point of the patients taking a 'longer look' to decide if they truly need the costly inject able or can 'survive' with other less costly therapies."

TABLE 1: The Five Steps of Biotechnology Management

1.	Procure and Schedule	• Commercial reimbursement (AWP-14.21%) equivalent to Medicare rate • MMA provides cover and ensures consistency across payers
2.	Area of expertise in Pharmacy Self-administration and compensation changes	• 76% of payers report using SPP, up from 64%· • 31% of SPP arrangements are semi mandatory • 78% of payers seek to drive self-administration
3.	Grouping Contraction and Chosen Products	• Preferred status in some crowded categories • Contracts for factor VIII, growth hormone, IVIG· • Rapid growth in MS, RA necessitates action
4	Grouping Contraction and Chosen Prod	• Differentials expanding between tiers· • Coinsurance still rare • Average nontiered copayment now exceeds $40 Out- of - pocket maximums rising
5	Advantage Intend and Cost Distribution Innovation	• Currently limited to antivirals, oncology, and rare diseases. • HER2/ neu • Gefitinib (Iressa) as first-line therapy in NSCLC subpopulation?· Physician/system readiness?

AWP= average wholesale price, IVIG = intravenous immunoglobulin, MMA= Medicare modernization act, MS= multiple sclerosis, NSCLC= non small cell lung cancer, RA= rheumatoid arthritis, SPP= specialty pharmacy provider.

Herein lies the challenge for the future. As costs rise, payers will turn to move innovative benefit designs that increase patient cost sharing burdens and provide employers with new coverage options (Vogenberg 2004a, 2004b). Payers consistently prioritize therapies that save lives, followed by those that spare disability. Other therapies, irrespective of the innovative nature of their technology or the unmet medical need in the category, will be associated with more difficulties relative to providing justification for high prices. Biologic therapies for plaque psoriasis, for example, offer better outcomes for patients, but they are prescribed for a condition that is not life threatening and at costs so high as to make payers balk.

In other words, in a world of scarce resources, payers give the advantage to life-saving therapies. Other products face significant challenges with respect to communicating value that is sufficient to guarantee advantageous access.

Patients' willingness to pay increasingly must be integrated into market planning as payers pass more costs to patients. Evidence has shown not only an overall fall in utilization at higher cost sharing levels but, increasingly, outcomes losses (Goldman 2004, Ozminkowski 2004). For payers, this raises difficult questions. For manufacturers, it means determining how to obtain favorable positioning relative to competing therapies and in absolute terms. Solo exclusive status means dramatically less when patients face a 25 percent coinsurance regime, particularly for non-life saving or non-disability-sparing therapies. Similarly, preferred status has little value if the non-preferred therapy has a comparable co payment.

A number of payers expect to introduce new benefit designs for specialty therapies in the next 6 to 18 months. Many already filed the necessary paperwork with state insurance commissioners and began selling the new designs in the market late last year so that the new models would take effect this year. Under these new models, intra-tier differentials widen, and coinsurance may apply to no preferred therapies. Although significant and difficult questions remain about the equity of these design- cost-sharing and, in particular, coinsurance, and highly7 regressive- they inevitably place greater pressure on manufacturers to demonstrate tangible (*e.g.*, life-saving or disability-sparing) effect, to trade price for access, or to try some combination of both.

Genomic Diagnostic Genetic Risk Management

At a meeting in 2002, payers discussing the management of TNF-α inhibitors agreed that while the available agents might work differently in different patients, this theoretical allelic heterogeneity did not necessitate coverage of every available therapy (Baker 2003). In addition, payers indicate that the burden of proof actually lies with the manufacturers. If a case can be made for one agent over another at the molecular or genetic level, then the manufacturers must demonstrate strong supportive data; only 14 percent of payers believe "allelic heterogeneity necessitates keeping all biologic agents in a class on formulary."

On the other hand, payers cite individual products for which highly specific molecular diagnostics exist and to which access in contingent on a positive test. Perhaps most familiar is the fluorescence in Situ Hybridization (FISH) test for HER2+ met static breast cancer, a positive result from which results in trastuzumab (Herceptin) therapy. While still relatively uncommon outside of oncology, virology, and a handful of rare diseases (*e.g.*, cystic fibrosis), the flurry of merger and acquisition activity within the molecular diagnostic industry reinforces the importance of the trend.

Payers have raised questions about the overall value of costly molecular diagnostics, particularly where incumbent-testing technologies may be more cost-effective (Farkas 2004). Faced with therapies that run into tens of thousands of dollars annually, molecular diagnostics increasingly represents a critical cost-saving strategy for payers.

For manufacturers, molecular gate keeping will create new challenges, as well as greater incentives to harness the promise of pharmacognomics. While some manufacturers view pharmacogenomics skeptically, fearing that it will unnecessarily limit sales growth, the converse may be closer to the truth. For example, although a large clinical trial failed to show that gefitinib (Iressa) confers a survival benefit for patients with non-small cell lung cancer at the population level, recently published data indicate that patients with an extremely specific gain of function mutation may actually benefit from earlier, more aggressive use of the therapy (Lynch 2004). Additional analysis has identified a district subpopulation of likely gefitinib responders (Pao 2004), and suggests that specific somatic mutations in the tyrosine kinase domain of the epidermal growth factor receptor may predict sensitivity to gefitinib (Paez 2004). In other words, molecular diagnostics could serve to increase gefitinib utilization in patients expressing the mutation, while simultaneously providing payers with greater control over access.

Bio-Pharmaceutical Value Chain

The bio-pharmaceutical value chain can be divided into; two parts, having as keystone the approval of the new drug by the dedicated public authorities such as FDA & IMEA.

In pre-approval activities, it concerns research & development. & in post approval activities it concerns the large scale production & marketing of a new drug, *i.e.* approval of new drug actually represents the boundary between cash absorption of cash generation.

Pre-approval Activities

Pre-approval activities are the starting point. of innovations in the pharmaceutical sector, which include the research of a new chemical or biological compound to be used for therapeutic purposes of development of the selected ones into new drugs which can be delivered in human being.

Pharmaceutical sector is a "research intensive" with the average R&D expenses that were more than 17.5% of the total revenue in the 90's & currently account for nearly 20% in the whole industry. The R&D expenses had been increased since the 90"s in this activity firstly it identifies a gene or motion or a sequence of both that is thought net step is target validation, which is concerned with a two fold study of the identified target.

Then it is necessary to define the interaction between the target and the whole human organism, in order to create a complete disease model, which allows scientist to understand the evolution pattern of the disease at the cellular level. On the other hand company has to check if there are some intellectual property lights on the selected target.

Before the study of biotechnology such type of activities did not exist, research processes were based on the chemical science and the study of the disease's symptoms. The same pattern showed for the target has to be followed by the chemical or biological compd. As genetic base of the disease evolution is known, the next step scientist has to do is to identify a set of compounds thought to have the designed effects in treating the selected disease.

Several different tests both in vitro and vivo, scientist select the compound that shows the best results, this is the compound, which actually represents the active principle of the future drug.

This final output of these research activities is therefore a drug prototype, which can actually enter the development phase that will test the efficiency and the safety of the new drugs.

The first development activity is the carrying out of preclinical tests, effects of the drug on animals.

Particularly, the mechanisms of absorption, distribution, metabolism, excretion and toxicology are studied in order to effectively trace the pattern of the new drug's compound inside the organism and make some connection before entering human tests. After completion of preclinical test positively; then the development phase of the new drug continues with the clinical test activity. But before entering clinical trials a first approval by the public authorities is required.

Clinical tests involve directly human patients and are divided into three steps, Phase I, Phase II and Phase III.

TABLE 2: Clinical Tests (*Source*: FDA, 2002)

Phase	*No. of patients*	*Length*	*Purpose*
1	20-80	Several months	Safely
2	100-300	Upto 2 years	Short term safety and effectiveness
3	1,000-3,000	1-4 years	Safety, desage and effectiveness

Phase I- Researchers test: New drug is given to a small group of people to evaluate safety. Mainly to determine safe dosage range, and to identify side effects, which include healthy participants and/ or patients.

Phase II- New drug is given to a larger group of people to evaluate effectiveness of the drug for the particular indication in patients with the disease and to determine the common short-term side effects and risks.

Phase III- Involves even larger groups of people to confirm the effectiveness of the new drug, to evaluate the drugs overall benefit risk relationship and to provide and adequate basis for physician labeling.

If all the above phase is successful then the specific public authority approves the new drug that can reach the market.

Post Approval Activities

Mainly consist of large-scale production and marketing of the new drug

They can be identified as

1. Purchasing
2. Production
3. Logistics
4. Marketing and sales and
5. Post-marketing test.

The technological and scientific base, are essential to reach the final goal of discovering and developing new drug.

Examples of the scientific and technological applications are geneomics and proteomics, which are the scientific study of genes and their role in the organisms. Body pharmacogenomics, deals with the specific differences in response of living structures due to the different genome of individuals and bioinformatics which is the creation, collection, storage and efficient utilization of biological data.

Time, Risk and Cost of the Drug Discovery and Development Process

In 90's use of biotechnology, the time to gain the approval for a new drug increased.

There are mainly two reasons why time is increased in R&D process.

In biotechnology there is treatment of more complex pathologies *e.g.* chronic and degeneration diseases require much more time in clinical tests for verification of effectiveness of therapies and biotechnology require constant revision of development process. Because of the weaknesses of rules concerting new biotechnological therapies the approval time is increased.

Another key feature of the drug discovery and development process is Risk. That is why only 250 out of 5,000 compounds identified in research actually enter the preclinical tests and only one out of five drugs passes the clinical tests.

TABLE 3: New Drug Approved Steps

Phase	*Completion time*
Target identification	1 year
Target Validation	2 year
Lead identification	4 year
Lead optimisation	2-7 years
Pre clinical test	1-6 years
Clinical test	7 years

So effect of both time and risk costs the R&D process is very high. Recent studies showed that these are increase in average cost of development of nearly 280% in the last decade.

Chapter 4

STRUCTURE OF BIOPHARMACEUTICAL INDUSTRY

Nomenclature

Two major typologies of biotechnology companies

1. Structure of biotechnology companies: Companies developing new products and 102 technologies related to the drug discovery and development process.
2. Complementary product/ service suppliers: Companies offering complementary products or support services.

Structure of Biotechnology Companies

Biotechnology industry can be divided into:

1. Product biotechnology
2. Drug agent biotechnology
3. Platform biotechnology

Product Biotechnology

They have objective of discovery, development, production and commercialization of therapentics products.

Drug Agent Biotechnology

They develop active principles for new drugs and give exploitation rights of their intermediate products to larger companies.

Platform Biotechnology

They focus on the development and commercialization of new technologies and devices both software and hardware, used in the drug discovery and development process.

Complementary Product/Service

Complementary product/ service suppliers can be divided into, service biotechnology and commodities biotechnology.

Service Biotechnology

They offer consultancy as well as support services to companies conducting the drug development process.

Commodities Biotech

They act as suppliers of specific chemical provisions, often developed accordingly with the instructions of pharmaceutical companies.

Analysis of typology of companies: for analysis of typology of pharma and biotech following aspects have been taken into account.

1. Business model
2. Positioning on the value chain.
3. List of top ten companies by market capitilisation worldwide level.

Big Pharma

Large pharmaceutical companies still rely upon an absolute market leadership position in the pharmaceutical sector.

The revolution in biotechnology weakness their competitive position as it strongly affected the competences required in the drug discovery process.

New technologies like the high through put screening technologies (HTS) and research approaches based on genomics and proteomics which allows the development of more effective drug with increased efficiency in R&D. Therefore big pharma had to change than business model.

Position of big pharma on the value chain

←——————————————————————→

Big pharmas

Drug → Drug development → Production → Marketing discovery

Big pharmas are downstream in the value chain for drug development, production and marketing whereas upstream in drug discovery.

Early phases of R&D are,

1. Research agreements with innovative biotechnology companies
2. Licensing agreements
3. Direct acquisition of small biotechnology companies' Source of the competitive advantages of big pharmas represents the strength in the post approval activities. They control the market through sales force and the garrison on distribution channels.

Product Biotech

They play a major role within biotechnology companies, with their direct action in the scientific innovation is transferred for these products have two major segments.

a) *Primary care segment:* Concerning the product actually and directly available for the end user.

b) *Secondary care segment:* Concerning the products that require the direct interaction between patients and specialized physician for their delivery *e.g.* new gene and stem cell therapy.

In the secondary care segment biotechnology had major impact as it allows entirely new therapies based upon genetic engineering of products *i.e.* biotechnology increases the efficiency of traditional drugs, introduces new concepts of drug product.

The position on the value chain shows how these companies carry out internally all the activities from basic research to marketing. Despite of high risk and investments in R&D reaching the market with a new drug is equal to cross the border between cash absorption and cash generation. The average positive profitability is a key characteristic of product biotech in comparison to the other kinds of biotech companies' analyzed. Such business model referred as ``virtuous`` circle in which products already on the market fund the research projects, thus contributing to generate a thick stream of potential new drugs through the pipeline and to enhance the long-term profitability.

To enhance the process of product biotech

A. Differentiate their product portfolio in personalized medicines, which maintains a competitive advantage on traditional large pharmas.

B. Co-operate with platform biotech *i.e.* to develop more efficient and effective technologies, which reduces risk in the drug discovery process, and also to implement diagnostic devices for personalized medicines.

C. Sustain their growth through acquisition of smaller companies.

Drug Agent Biotech

They carry out drug discovery and development process up to the lead identification and validation *i.e.* output represents on intermediate product requires a further development phase.

Drug agent biotech operates in the industrial market and has as customers' larger companies completing the development process and if successful, marketing the drug.

Drug agent biotech's current business model is based on generating returns from the licensing agreements, which occur after lead optimization. But in some cases the development process of the new drug is carried out up to the clinical test, and products reach the higher transactional value before commercialization. Therefore drug agent biotech covers only part of the value chain.

Make use of internalities in their product portfolio, searching for possible side applications in fields other than the pharmaceutical, which increases the amount of royalties related to a single product.

Diversify their business in platform technologies development thus coupling increased efficiency in internal research with increased revenues from supply to other companies. Position of drug agent biotech on the value chain.

Drug agent biotech

←——————————————————→

Drug discovery→ drug development→ Production→ Marketing

Platform Biotech

It represent a complex set of companies based on different technologies which operates for one biotech companies their role in this sector can be analyze by further distinction of products between:

1. *Process technologies:* Improves the efficiency and effectiveness of the drug discovery and development process

2. *Product technologies:* Which are embedded in the product as *e.g.* new delivery system or associated with product as the diagnostic tools

This leads to define four typologies of platform biotech.

1. X- omics platform
2. Bioinformatics
3. Drug delivery platforms
4. Diagnostics platforms.

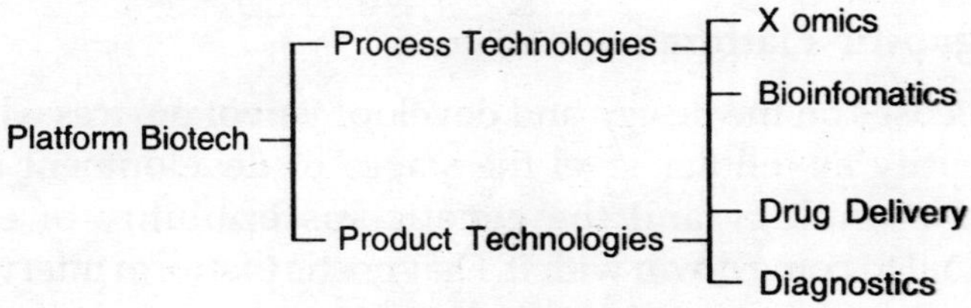

Fig. 2: Platform Biotech

1. X-omics Platforms:

It intervenes in the basic research activities, developing and commercializing technological tools and devices to support the drug discovery process. Examples of this are genomics and proteomics platforms that develop systems for the structural and functional analysis of genes and proteins as triggers of the cellular mechanisms. Along with this there are other relevant platform technologies, which are under development.

a. *Phenomics:* Analyses structure and visible characteristics of microorganisms.
b. *Metabolomics:* Analyses the effects of on the cellular mechanisms of eternal stimuli or of a genetic mutation.

2. Bio-informatics Platforms

It develops and implements software and hardware tools, to generate, collect and manage data in output by the drug discovery process. It act at different stages of the bio-pharmaceutical value chain, major contribution is in the creation of a knowledge base for the whole drug discovery and development process.

3. Drug Delivery Platform

It supports development and commercializing system and devices to effectively delivery active principles in human bodies. Biological molecules are more complex to manage in industrial processes than the traditional chemical ones. Their interactions with human bodies are difficult to control. To deliver the right drug at the right place, one should study all new improvements in the genetics. Drug delivery platform play their role in two sides, improving the characteristics of traditional delivery system and creating completely new system.

4. Diagnostic Platforms

It focuses on the design and development of devices which can identify at cellular level the stages of development of a defined pathology and the genetic susceptibility of each individual to come down with it. Diagnostic platform intervene during clinical tests and if the drug is successful, then their products are commercialized in the final market as a complement to drugs.

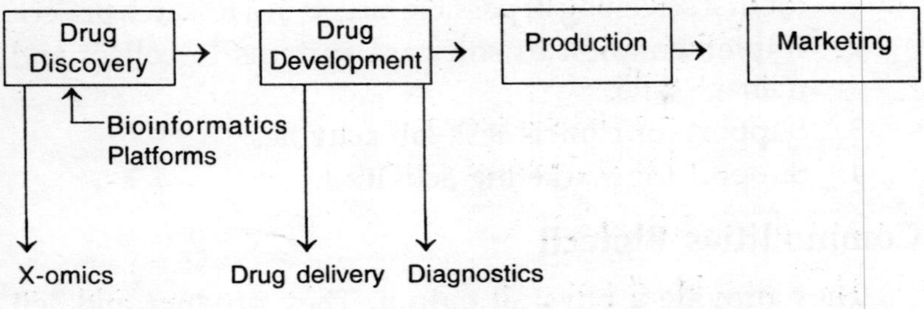

Fig. 3: Position of Platform Biotech on the value chain.

Platform biotech has a time to market and a failure rate consistently lower than the drug oriented companies.

Business model for platform biotech have,

1. To develop customized product, which can fit the needs of drug oriented companies, maximizing efficiency and effectiveness in different phases of the value chain and in different therapeutic areas:
2. To force the creation of a technological standard by allowing companies to complete on the whole market and avoiding in house solutions from core biotech companies.

Platform biotech is not yet profitable but in comparison with drug agent biotech, they present three time's higher revenues and market capitalization. They are unable to generate a sustain long term profits; and will not reach the results of top product companies with their pure technologically focused business model.

Service Biotech Company

It acts as external source of information and competence particularly for drug oriented biotech companies.

Major services in biotech companies:

1. Collection, of biochemical and molecular data from public and private sources. This helps biotech companies for first screening of possible targets in a research project.
2. Implementation of software systems to collect and manage data.
3. Support for clinical test, lab activities.
4. Support for marketing activities.

Commodities Biotech

They provide a physical output. They produce and sell consumable products, for drug research and development process. These products have high degree of specialization. But these products give a marginal contribution to the overall value of the final products.

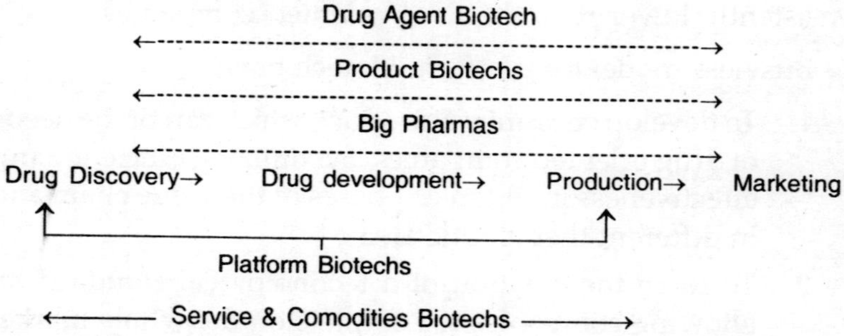

Fig. 4: The Position of Bio Pharmaceutical Companies on the value chain

Biotech industry structure: The US is commonly known as homeland of biotechnology because of following three points- a. Historical, b. Financial, c. Normative.

TABLE 4: Global Biotechnology Industry Data

Area	*No. of companies*	*No. of employees*
Europe	1,878	87,182
USA	1,466	142,900
Canada	493	7,785
Asia	601	9,764

Chapter 5

ORGANIZATION, MANAGEMENT AND STRUCTURE

General Management Composition

The new management challenges changes in styles and amounts of funding, the rate of change in information transfer opportunities, new and different relationships with partners, changing patterns of employment and increasing uncertainty will all contribute significantly.

Panel of Trustees

The Panel of Trustees is an assembly of highly competent individuals ready to carry out their tasks with significance and perseverance. In the Panel there are qualified and capable leadership, and good equilibrium of both competencies and geographic representation. The Panel has to improve the gender equilibrium in its appointees, as well as to gain specific competencies, such as legal proficiency in intellectual property rights, which will be important in the future.

Particularly those decisions concerning to the budget, the most important policy decisions the panel takes yearly. All Board members, irrespective of the specific committee on which they serve information on these aspects needs to be available to them. The flow of information from the committees to the full panel, and from management to the panel could be enhanced.

The Panel must be guaranteed that all outputs are costed completely and entirely, that the production of outputs is resourced adequately, and that they are delivered on time. This come within reach to manage the changes resulting from the enduring decline in unrestricted core funding.

Financial Management

This area has been one of the most demanding of all aspects of management.

The full project costing, showing both income and all expenditures, will be one of the most important ways that can manage the implications of the long-lasting decline in unrestricted core funding. It believes that the present categorization of funding (unrestricted core, restricted core, special project and self-generated income) is one aspect that places a high administrative cost on centers and complicates financial management.

Prepare all future budget proposals on the basis of full project costing, showing both income and all expenditures for project outputs.

There is still significant scope for improving the integration of "end of month" financial data transferred from regional posts to Headquarters, with the costs debited by staff at Headquarters to those regional specific activities.

The inclination for some managers to keep separate sets of financial records should be discouraged. When center will operate much better when a single database and the related sets of figures are, without exception, used by all managers.

Human Resources Management (HRM)

Noteworthy progress has been made with job classification studies and the review of positions for both internationally and nationally recruited staff. These initiatives have reduced anomalies and seek to ensure that center's terms and conditions are fair and rationally competitive with comparable enterprises. Progress is being made, within this context, on mechanisms for performance appraisal, again for both internationally and nationally recruited staff, and pilot teams have been identified to test new styles of multi-source performance appraisal.

The availability of professional HRM advice to having an impact in assisting programme directors to develop the necessary understanding and philosophy of people management that are needed to ensure center remains efficient and effective in a changing world.

There are opportunities for improved administrative efficiencies through closer interaction of systems between Human Resources and Financial Management, and the move to project-based management will further highlight the need for these improved systems.

Long-term initiatives have been commenced to change the work culture of the organization, staff and research partners, when planning collaborative work, seriously consider gender issues. Actions have been undertaken to identify potential leaders, provide training in both leadership and team building, and look for particular ways to express and encourage participatory management The areas of job classification and the terms and conditions of employment is particularly important for both national and international staff. The center should encourages to continue to look at a range of examples so as to ensure that what is being done is most appropriate for the future employment environment. Because of changing circumstances the future will require a number of changes compared with what organizations have regarded in the past as best practice.

A particular confront is that scientific research is strategic and long term. However, the whole funding environment has changed and the focus is on shorter term funding than was the case in the past. This will mean a variety of employment relationships, some of which will be quite different from the very long-term tenure that many enjoyed in the past. The job classification and reward systems initiated thus far may have to be amended to take this into account.

Attention will also need to be given to the feeling of "overload" being experienced by many staff and to providing effective cover for key persons, when taking parental leave, study leave, and when travelling.

Information Technology and Management

In any scientific organization information is probably the most expensive source. This includes the ideas members of staff have, and how those ideas are developed, tested and subsequently shared. The skills in managing information are therefore very important. The resources and opportunities for managing information have developed enormously since the mid 1980s with the availability of computers and related systems.

Having such a wealth of information poses a challenge as to how it is most effectively managed. There has been a good history of software development for specialist applications and a number of significant scientist-led initiatives.

Agricultural researchers face increasing challenges for the efficient management of research results. Biotechnology promises to generate vast sets of data on genetic variation, but breeders will need these data to be directly linked to crossing plans and field evaluations. Concomitantly, increasing costs of field and laboratory research mean that experimental data both old and new and from diverse sources - must be exploited to achieve greater cost-effectiveness, and that such data must be managed as valuable corporate resources. Although most centers and some NARS have explored use of research databases, there are few examples of fully integrated crop information systems that meet data management and storage standards of other industries.

Use of a standard interface across centers not only ensures development of more versatile and robust software but also will allow major savings in software development, training and system protection.

To guarantee continued efficient development, the initiative needs stable researching and a formal management structure.

Management Information System

Outside of what has been done in information management, that some of the technical aspects of developing and maintaining management in sequence systems have not received the overall attention they warrant. The short- and long-term cost to the Center is largely unrealized.

Management of Intellectual Property and Intellectual Property Rights

It is acknowledged that there are two initiatives that have to proceed simultaneously. Individual centers have to develop their own clear policies for the management of IPR and the situation needs to be avoided whereby individual centers have widely divergent policies.

Policy related to genetic resources held in trust, the products of research, and information products and computer software

Economics

Economics Programme (EP) provides valuable services to its internal partners and undertakes disciplinary research that is highly regarded by external collaborators. This assessment is based on the excellent documentation provided by the programme, discussions with its staff as well as staff of other programmes, internal reviews undertaken by the EP, and the report of the

Staffing and Budget

Irrespective of the location of the staff, the proper balance between support to regional activities and Headquarters-oriented work needs to be monitored constantly.

The staff of the EP, is concerned about being overloaded, partly due to what they perceive as an increase in administrative, coordinating, and reporting activities, many of which are unavoidable. What this strongly suggests is a need for better priority setting and time management, both by the staff and their managers.

Programme Activities

In general, Economics Programme seeks to develop the economic benefits resulting from the scientific work of the Center, and also addresses external constraints to the adoption of improved technologies and efficient utilization of agricultural resources in different crops and regions. The first group of considerations has led to substantial work on priority setting, while the second has led to impact assessment activities. Two other considerations have been taken into account while designing the work programme, *i.e.*, ex ante evaluation of technologies with high potential, and second, the role of the private sector in the extension and use of new technologies.

The EP strives to serve several internal and external partners mainly management, the commodity programmes. From all indications, the EP has served management well it assists the Director General in reporting on the Center's achievements and impact.

Support to Commodity Programmes. The Economics Programme has had a long history of supporting the commodity programmes in the areas of technology design and impact assessment. It needs to take a more proactive role in priority setting in the commodity programmes, which are looking for more detailed guidelines, as well as a clearer basis for assigning relative weights to different research objectives, such as breeding maize for food or feed. The issue of popularizing hybrid versus open-pollinated maize, say, in Africa, also needs specific answers. The Panel considers that the findings of the impact studies should also feed back into the appropriate programmes, and result in suitable changes in the design of commodity research projects. For example, the Panel is convinced that the EP's findings on the socioeconomic reasons for the non-adoption of improved varieties of maize in over 40% of the target area should directly contribute to the collaborative work between the two programmes on this important issue. Another example of the desirability of greater inter-programme collaboration is the frontier project FP6, Priority Setting and Technological Forecasting for Increased Research Efficiency proposed to be carried out exclusively by economists from the EP. It is clear to the Panel that greater concerted efforts by staff of the EP as well as other programmes are needed to further integrate the commodity programmes into this type of research project.

EP staffs in the outreach programmes perform a variety of very valuable functions. For instance, they organize training programmes, mainly in the form of short-term workshops. The themes selected for such workshops in recent years include natural resource policy, economic analysis, and seed production and supply policy. Currently, training programmes on these topics are organized in Africa and Central America, and the staffs are involved in collaborative training programmes with the commodity programmes. It is expected that staff of the EP will also make significant contributions to the proposed training course on sustainable systems. However, no such training programme is presently conducted in Asia.

Precedence Setting

National economies, developed as well as developing, are getting more and more integrated. Institutional mechanisms such as WTO at the global level, and the regional economic groupings at the regional level, are leading towards the opening up of national economies and their agricultural sectors to international competition. Such moves have two important implications for the global agricultural agenda. First, these developments more sharply highlight the question of comparative advantages (or disadvantages) that a country faces in the production of certain commodities. Research priorities have to take into account, in a dynamic and fast-changing setting, the comparative advantage of growing a particular crop in a given region. Second, the opening up of the economies may lead to the "import" of price instability into a country. At the same time it also provides, via the trade routes, an opportunity to stabilize supplies and prices. Thus, a whole range of new policy issues connected with "stability" acquires importance in research of the EP.

Together with the move towards global economic integration, there is also a pressure to liberalize economies, including their agricultural sector, by removing state interference in the production, pricing, and trade of commodities. A rapid shift to a market orientation is clearly discernible. Often, this is interpreted as withdrawal of the state from activities that can best be managed by the private sector. Agriculture is no exception to this general tendency. While research is still considered a "public good", justifying a major role for the state or state-supported institutions, whole ranges of activities, from extension to input supplies and especially marketing, are now considered as the legitimate areas for private sector operations. The EP's study on maize seeds has explored this area; but privatization does not necessarily means opening the market to large seed companies. The potentialities of the smaller, national companies and the constraints that they are facing, should, in the Panel's view, be a legitimate topic for research. Such research would also need to link up with macro-policy aspects, such as prices and trade policies.

In recent years, new frontiers have opened up in two broad areas of science, which are of immediate relevance to the Economics Programme. Initially, the growing importance of biotechnology, and secondly, significant developments in the area of information technology. Developments also have important implications for the Economics Programme. Questions such as the ex ante evaluation of new technologies, priority setting in an unfamiliar environment, forecasting the adoption of largely untested technologies, etc., have thrown up difficult challenges to social scientists. A much closer interaction between the social scientists and biological scientists, as well as with the user systems, has become imperative. The rapid advances in information technology, on the other hand, provide a facilitating environment for dealing with the conventional issues as well as new developments in the social sciences. Such developments require not only the acquisition of new skills, but also a change in the mind-set to comprehend and adapt to the emerging scenarios.

Demand and Supply for Commodities

With the increase in incomes and the decline in the population growth rate the world over; the rate of growth of demand for cereals, including wheat and maize, is decelerating. Still, there is hunger and poverty in large parts of the world; and the switch-over from coarse cereals (excluding maize), roots, and tubers to superior cereals, *i.e.*, wheat and maize, is taking place in the initial stages of economic growth. Yet the demand for maize and wheat as food is not increasing as rapidly as in the past, while the demand for animal feed and for industrial uses of these cereals is progressively increasing. Even in their food intake, consumers are exercising more product differentiation. These developments have immediate implications for the projection of demand and supply.

Administration Configuration for a Project

The administration configuration is very important for a project the coordinator of project has responsibility for establishing the network and achieving its objectives, including project reporting and accounting. A project manager, whose role is to provide day-to-day support and implementation of all management and organisational tasks, supports the coordinator. A Steering Committee, chaired by the Coordinator, oversees the work of the network and assists the coordinator in the management of project. The Steering Committee meets at least quarterly.

In addition to the Steering Committee, a Work Package Coordinating Committee and a Training Committee oversee the progress of the work packages and training activities respectively.

Chapter 6

ESTABLISHMENT OF FOUNDATION OF BIOTECH INDUSTRIES

The establishment of foundation of biotech companies based on new idea about new pharmaceutical compounds or technological devices. In many cases these new companies are the outcome of spin-off processes. There are two types of spin-off processes:

Academic spin-offs, in which the universities or the public research centers play a major role. It starts in the academic labs and a new firm is created by university people in order to exploit it commercially;

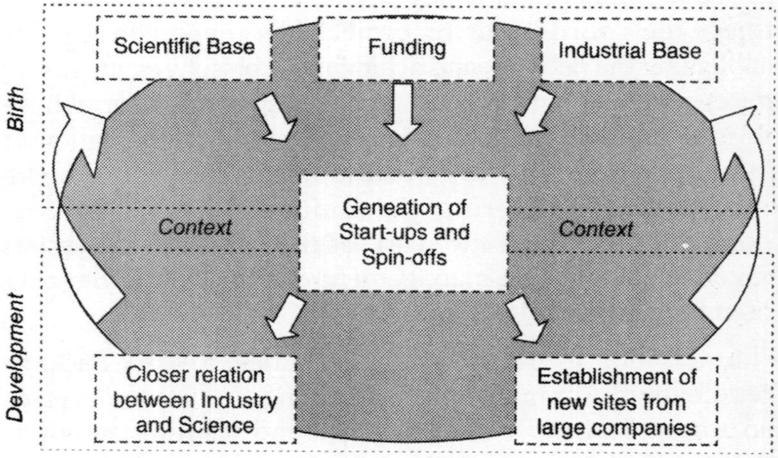

Fig. 5 : Growth Mechanisms of Biotech Industry

Industrial spin-offs from large pharmaceutical companies. These spin-offs are planned by the corporate, which facilitates the creation of independent firms focused on biotech businesses.

The process of generating new companies also requires the availability of funding programmes tailored to the funding of new high-tech ventures.

This is increasingly true in biotech, where new biotech companies require large amounts of money·from the beginning. A common distinction is made between: product-oriented biotech, whose objective is to discover and develop new drugs; and platform-oriented biotech, which are aimed at developing technologies (Genomic, proteomic...) that support the research process. The former faces a higher risk and a pay oriented business models are more concerned with a shorter term and lower risk investment. In both cases, the technological and scientific improvements made by biotech-related disciplines and technologies do not more allow "craft-made" solutions": the creation of a new biotech firm currently needs more than US$1 million on the average. Hence, availability of funds is a third key factor in a biotech industry.

Finally, a fourth factor is given by the characteristics of the general context: the presence of a favorable "environment" (normative, social, historical, and infrastructural) can actually facilitate the birth and the development of biotech industry.

Once the process is started, a virtuous cycle often begins. The strong presence of new innovative biotech companies increases the area attractiveness, facilitating the establishment of new sites (particularly research sites) from large biotech or pharmaceutical companies. The academic origin of some companies, moreover, facilitates the establishment of strong links and networks between industry and science. These two effects, in turn, reinforce the industrial and the scientific base of the area and therefore provide the basis of the generation of new ventures (which becomes a factor of attractiveness of financial actors) and so on.

Driving Force and Practices :-

Four main driving forces can be identified, on the basis of the analysis done.

- Financial driving forces, which concern the availability of funds for the biotech companies;
- Scientific driving forces, which concern the exploitation mechanisms of scientific research;
- Industrial driving forces, which concern the exploitation mechanisms of industrial research;
- Supporting driving forces, which concern the presence of a favorable general context.

For each driving force, a set of factors that enables the driving force to positively act on the area for the birth and development of industry is identified.

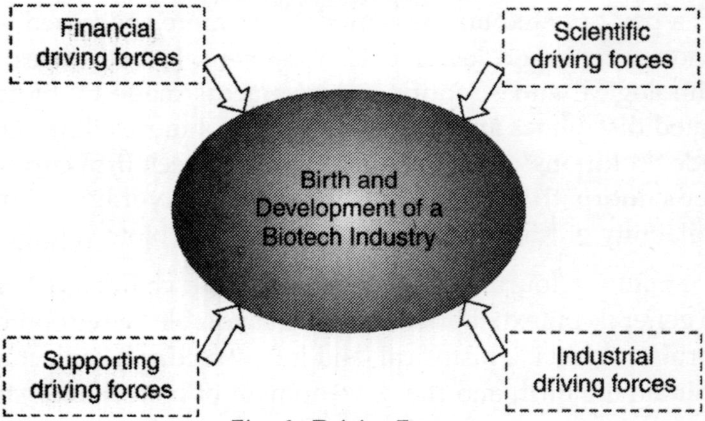

Fig. 6: Driving Forces

This set of factors is split into categories:

- Actionable factors:Which can be directly managed throughout actions made by the actors of biotech industry.
- Context factors:Which concern primarily the historical, social and economic background of the geographic area and can not be reasonably modified in the short and middle term

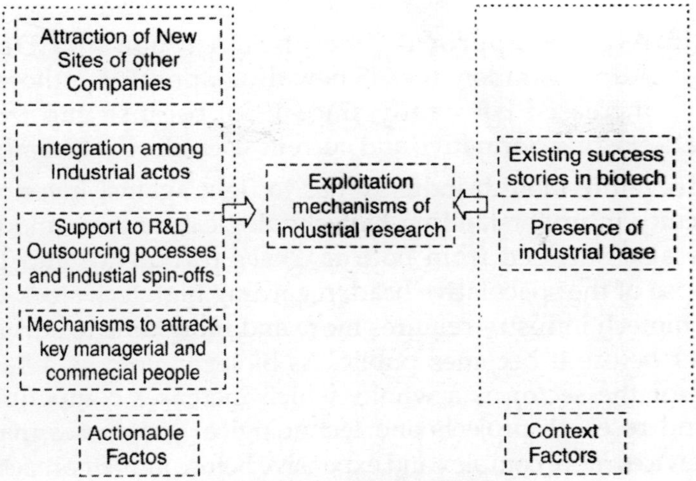

Fig. 7: Financial Driving Forces

Financial Driving Force

Biotechnology is widely recognized as the healthcare paradigm of the future, which changes the traditional pharmaceutical chemistry. The scientific and technological advances in bio-technology drives down costs and reduces time needs by increasing productivity, and strengthen returns by giving more effective therapeutics with fewer side effects.

In spite of developments in this area still the process of developing drug is long risky and expensive.

Recent data (Tuff University, 2002) suggest that the average cost of a new drug has risen to US $802 million and often requires 14 to 15 yrs to reach the market. Only 250 out of 5,000 screened compounds can enter the preclinical testing. And only one out of the five drugs that enter clinical trials will be finally approved. There are so many causes for this but some are,

(i) Biotechnology allows to treat more complex pathologies (*e.g.* chronic and degenerative diseases), which are less known; because of this it has higher failure rates and longer clinical trial.

(ii) Main technologies are still in developing stage far from maturity stage; therefore they require more investment of achieve state of art devices.

(iii) Average approval time of FDA (Food and Drug Administration, the US new drug approval authority) increased constantly since 1998, because gap from existing normative and current scientific environment.

Therefore biotech industry has to fight against investor's difficulty in understanding biotechnological business model. They also suffered from both market's turbulence and the collapse of the speculative bead; regarding high tech stocks. A new biotech industry requires more and more time to survive longer before it becomes public. As biotechnology is a basic skill for the sector as a whole, which increases competition around research projects and technological progresses make lab devices more complex and expensive before industry reaches the capital market, the main way to raise capital are private equity investors, especially venture capitalists. (VCS). VCS are fund managers who invest money into private companies in exchange for an equity position. At an earlier stage of development, writing a complete and successful business plan, being able to catch the attention of the business community and to demonstrate the commercial exploitation of a scientific idea, requires approx US $15,000-50,000.

The main problem of an investor is the way he can disinvest his funds gaining a profit. The most important exit strategy is IPO but recently it has become quite hand to perform, making private sectors more careful. There are only three significant equity-financing windows. Because of Sept 11th attack in 2001 and finally the explosion of Iraq war in 2003, all made financing decisions more difficult.

In comparison to 2000 the financial resources available for the US sector are reduced to less than one third in 2002. Private firms can raised only US $1.1 Billion as compared to US $3.7 and 3.9 billion in the two previous years. Recent studies (Di masi 2001) suggest that almost 35% of pharmaceutical research failures are due to financial problems.

The analysis shows that the funding of new biotech industry is rather complex. It should concern the whole life of the start up from the foundation to the listing on the stock exchange and require a number of financial instruments.

There are five factors can be identified corresponding to different stages of the industry's life cycle.

1. Availability of pre-seed capital.
2. Availability of seed capital.
3. Availability of venture capital.
4. Availability of government funds.
5. Availability of exit strategies.

Availability of Pre-seed Capital

This is the capital, which can be, used for start up carryout a proof of concept work and develops a credible business plan. (Approx US $15,000-50,000)

Some practices, which can enhance the availability of pre-seed capital, are as follows.

1. To develop local pre-seed funding and to fund scientist in writing business plans. The traditional scientific education does not provide as managerial competences as needed in order to start new venture. The preparations of a credible business plan represent a major problem for entrepreneurs, who want to gain access to public and private capital. After a first screening of the scientific ideas, pre-seed capitals managed primarily by universities a technology transfer offices linked to universities are assigned to the most brilliant scientists to help them purchase the consulting services needed to transform their scientific ideas into business ideas.

2. To create support services like consulting offices within universities. These offices should be able to evaluate the potential for business exploitation of scientific ideas and hence provide managerial and legal competences to plan the creation of new start-ups. As universities cover wide range of disciplines they could exploit synergies among different departments, stressing the interaction between science oriented and management oriented academic people. Instead of directly funding entrepreneurs the same goal is to make them able to write a business plan that is reached through the inside offering of dedicated and complementary competences.

3. To provide early funds in competition. Which gain the two-fold effect of stimulating the entrepreneurship culture among scientists and of selecting the ideas to fund after the completion of business plans. Given the amount of funds available in these competitors incubation platform a programme of the Genepole of Evry has mission to provide pre-seed capital to identify potential company directors from public or university research laboratories or from private companies and help them bring their project to friction. They provide the new entrepreneurs with management training and basic information on industrial property, safety, and regulatory standards. In the pre creation period the incubator covers the cost of forcing the team of external service providers and of a marketing strategy consultant; as well as helping entrepreneurs to search for early stage financing.

The GSAS Havard Biotechnology club a profit organization with more than 4,000 members' most events and provides services to explore the world of business and biotechnology operates inside the Harard University. The mission of club is to bridge the gap between industry and academia by building relationships with companies operating in the biotechnology. They hold business plan competition for biotech related companies each year.

Availability of Seed Capital

This is the capital with which a new biotech industry can start up (app. Less than US$ 1million). The main source of seed capital is from business and individual who invest in private companies, taking over the risk of long-term business in exchange for an equity position.

Some practices which can enhance the availability of seed capital.

1. Development of local seed funding for biotech industry start-ups. Universities or local organizations with sponsorship of private companies, banks, etc can directly manage these.

2. Support for investors evaluate biotechnology business model. Drug oriented companies' face a higher risk and a larger

pay back time than technology oriented industries but equally have bigger expected revenues. Biotechnology related disciplines and their degree of specialization make the evaluation of new ventures by investors more difficult. External and qualified support from universities or public or private association may help investors in reducing the informative asymmetry and evaluating the scientific reliability of the business ideas. The consulting offices provide such kind of support.

3. Creation of a network of potential early investors. Major problem in the development of a business idea is the lack of information about funding opportunities Genopole ler Jour fund is a seed capital fund, which has started in oct. 1999. Managed by the Genopole of Evry. They directly invest in biotech industry. Fifteen private and public investors including a balanced mix of business, banking and institutional actors have subscribed to the Genopole ler Jour fund for €1.2 million. The evaluation of fund takes into account the work and know how of the creators and previously financed projects. The time span for the contribution to be given back is usually 8 years. This initial capital leads to public financial assistance in particular provided by ANVAR, the French innovation agency; through innovation aid programme, that acts as the financial leverage for the next development stages. Some successful cases of financed firms are monoclonal autibodies theraputics (MAT) and Obe therapy biotechnology.

San Diego Tech Coast Angel is a network of private investors who invest in and assist early stage southern California companies. It is a part of a larger network named southern California based Tech Coast Angels organization; which also has network in Los Angeles and Orange County and has invested over US $ 250,000- $2 million, this investment range is generally not of interest for venture capital fund. The goal of this network is to allow biotech industry start-ups to launch their business and to accomplish critical milestones that will make them attractive for larger venture capital financing.

CC&D CONNECT founded in 1985 at the urging of San Diego's business community is the globally recognized university based public benefits organization fostering entrepreneurship in the San Diego region that catalyses accelerates and supports the growth of the most promising technology and life sciences businesses.

Part of the University of California, San Diego (UCSD) CONNECT has a dual role in accelerating growth: which provides added value and delivers targeted, high level expertise to San Diego's technology business community by teaming up with the region's most prominent industry specific organizations and individuals and by partnering with world class UCSD resources such as the school of medicine, Jacobs school of Engineering, San Diego super computer centre and Scripps and Salk institutes.

Availability of Venture Capital

Venture capitalists are fund mangers. They invest into private companies. In 90's venture capital funds became larger and their minimum deal size has correspondingly increased to the point where they currently does not consider financing of less than US $ 3-5 million, venture capitalists who fund biotech industries have a strong experience in the industry and an established track records helping start ups to become mature operating companies. They have networks and therefore can help the new entrepreneurs to recruit employees, executives, directors but also customers and others investors. They negotiate for a large equity position and retain the managerial control of the industry whereas the founders are free to develop their scientific idea.

Some practices which can enhance the availability of pre seed capital.

1. To facilitate the access of companies to international funding networks, networking attitude of venture capitalists is more evident, expanding the funding possibilities for the new companies.

2. To attract acknowledged biotech-oriented venture capitalists. Venture capitalists are the last step before access to the capital market. Current global recession, particularly for the high tech sectors, enhances the importance for the companies of having a well acknowledged biotech oriented venture capitalist among their major shareholders. UCSD CONNECT'S financial forum traditionally attracts, venture capitalists, pharmaceutical and biotechnology executives, investment bankers, private investors and service providers. The programme combines CONNECT'S 12 years of biotechnology corporate partnering experience with 18 years of financial forum expertise; both have played an important role in the development of San Diego's start up biotechnology and pharmaceutical industries since 1985. It provides a showcase for San Diego's most innovative bioscience companies and researchers.

For this industries/ companies have to satisfy three main criteria:

1. They had to have already received venture capital or significant seed investment.
2. They must be in need of venture funding.
3. They must be related to the life science industries, *i.e.* discovery instrumentation and/ or software and bioinformatics.

Availability of Governmental Funds

It represents direct intervention of the local or national government in funding biotech industries. Government funds facilitate the fundraising of industries from private investors, enhancing the industries credibility. Public funds can be available for all the developmental stages, but main problem of such kind of funds is that in most cases the small starting ups fail in encoding the bureaucratic mechanisms. Biotech industries in early phase of developments need a great amount of cash and do not generate revenues, generally cannot wait for public funds and are forced to search for public funds and are forced to search for complementary sources of funds. In order to enhance the availability of governmental funds and to make it more suitable for biotech industries the following practices can be highlighted.

1. To develop governmental VC style fund. The VC style mean less bureaucratic approach and to a closer interaction that is not limited only to the funding process but includes support activities and even managerial support.

2. To provide specialized biotech funding programs. In 1981 the general assembly of North Carolina created an organization to stimulate the development of biotechnology known the North Carolina biotechnology centre.

Initially it was founded as a state government body, but then centre was reconstituted in 1984, as a private, non-profit corporation, giving it greater flexibility and the role of catalyzing interactions between industry, academia and government for technological development.

The Bio Regio Contest has been most powerful organization in development of biotech industry in Germany. The federal government in 1985 launches this; its fundamental objective is to establish strong linkage between basic research and commercialization of biotech activities. For this purpose federal government made available € 25 million in these regions that demonstrated to be in strong competitative situation and to have great prospective of growth in biotech.

Availability of Exit Strategies for Investors

The main objectives of the investors *i.e.* both private and public, is to remove their funds gaining profits in the mid term horizon, by selling of their equity position in the funded companies. For this they sell the equity position to other investor and finally to the capital market.

For this, most important exit strategy is IPO Represents for the company the only way to sustain its autonomous growth in the long term, And for the early investors the easiest way to gain profit from their shares.

Capital markets dedicated to high growth stocks, *e.g.* NASDAQ in the US enhanced the possibility for high tech industries to reach the retail investors and to enlarge the set of their shareholders, offering a financial environment more suitable to the businesses of such kind of industries, And creating a customer base for the industries stocks more oriented to high risk and long pay back time than the available in the traditional markets. For creating a dedicated capital market it requires long period of time and strong consensus by the central government and a strong commitment of the financial community: both factors go beyond the range of activities that can be undertaken to support the creation of a biotech industry.

Scientific Driving Forces for Biotech Industry

Five factors can be identified, which influence birth and growth process of biotech industry.

1. Presence of scientific base.
2. Technologies transfer mechanisms.
3. Networking culture.
4. Entrepreneurial culture.
5. Mechanisms to attract key scientific people

The creation of a scientific base is part of the history of the area and generally goes beyond the range of activities that can be undertaken to support the birth and development of a biotech industry since it is context factor rather than an actionable during force.

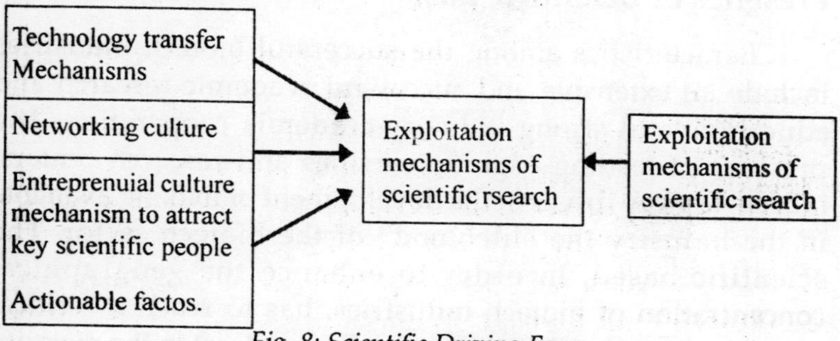

Fig. 8: Scientific Driving Force

A technology transfer mechanism allows scientists to commercially exploit the results of their researches through the licensing to existing industries or allowing processes with state of the art technologies and procedures. Their technology transfer offices universities and research centers can leverage academic knowledge and make it more easily accessible to industries. The commercial exploitation of academic knowledge enhances the reputation and the competitive position of the scientific base. Universities can access new sources of funding as well as other forms of pay offs and are forced to set up internal mechanisms within their organization fostering the successful commercialization of their research results.

The development of technology transfer offices make use of the networking culture among scientists and their outward looking at industrial needs, favoring both formal and informal exchange of knowledge and competences with close relationship with industrial bases there is a diffusion of the entrepreneurial culture among scientists.

As they are open to industrial stimuli, academics can understand industries real needs, the potentials and modalities through which academic research results can be exploited. Such diffusion of an entrepreneurial attitude within universities and research centers reaches themselves in the creation of new industry. By creating an excellent and innovative scientific environment one can attract scientific people to this area.

Presence of Scientific Base

Characteristics among the successful biotech industries include an extensive and successful academic research and education, and strong industry-academia cooperation. The presence of leading-edge universities and research centers, indeed, is a key driver in the development of making available in the industry the "lifeblood" of the biotech sector. The scientific based, in order to enhance the geographical concentration of biotech industries, has to reach a "critical mass" in the research activities in all (or at least in the majority of) the biotech-related scientific fields. To create a scientific base can be carried out only by a public actor (typically central government) with a long-term time horizon and requires huge investments for the infrastructures and the human resources. Even if infrastructures can be built in relatively short time, it is quite impossible to establish ex-abrupt the excellence in science, therefore, as noted; the presence of a strong scientific base is the result of the "history" of the area, thus representing a context factor in our analysis.

Technology Transfer Mechanisms

Leading edge research is a primary driver to any industry, but unless this research can be effectively transferred to the marketplace, the benefit to the regional economy is limited. Technology transfer is the process of finding, creating, and leveraging whether through licensing or the creation of new products—intellectual property that has potential commercial applications. Such applications are the fruits of research conducted within a variety of research universities and institutions. In the biotech sector, relevant "sciencetification" processes, most of research results are easily used by industries as "component/products which are ready for use" and not as "raw materials" which must undergo further long transformation phases. A new discovery in genomics, for example, is directly usable by bio-pharmaceutical companies. On the one hand, "translating" lab researches into products or technologies may be performed through the direct creation of academic spin-offs.

To develop efficient and effective technology transfer mechanisms, the following practices can be highlighted:

1. To create dedicated technology transfer offices. These structures, generated within universities, should have a clear managerial independence and maintain a flexible and "thin" organization.
2. To enlarge the range of activities of existing technology transfer offices, offering a complex set of supporting services, from the general and project management "consultancy" to the access to both public and private funds.

3. To conduct formal analysis to better understands the creation and flow of IP involving faculty, researchers, and students. Local technology transfer offices should be tasked to produce explicit recommendations for the adoption of new IP management-related processes.
4. To "match" innovation demand and offer through a stronger involvement of researchers in commercial objectives, "driving" in part the research activity towards the development of useful industrial applications. A strong support is needed for existing networks, both formal and informal, between industry and academia.

The UCSD Translational medicine Program (TransMed) has been created to assist scientists in the critical task of moving medical research closer to commercial ready medical technology within the University in order to benefit patients and the public at large. This program is specially designed to support promising "translational" research, which is still laboratory based, but is nearing the stage of clinical testing and application.

Networking Culture

Networking culture refers to the ability to create closer relationships within universities and research centers and between these ones and existing industries. Relations are favored by localization, but the process of formal and informal networking may also be strengthened through direct actions.

To help the diffusion of the networking culture among researchers, the following practices can be highlighted:

1. To support cooperative projects among actors within the industry cases this allows people from industry to "ripen" products or processes in a close proximity to scientists, whose inputs are useful for further development. Cooperative projects offer the possibility of linking technology, capital and know how to accelerate the technology commercialization, eventually nurturing new knowledge based ventures;

2. To promote public and informal meetings. It is events or conferences where relevant biotech companies and academics are encouraged to interact effectively with each other.
3. To create an environment favorable to networking, key task in enhancing the diffusion of the networking culture is the establishment of a 'positive thinking" among actor. Actions may be taken in order to enhance credibility and trust among actors.

Bay Bio's Major Objectives are as Follows:

1. To expand communications within the bioscience community;
2. To increase the public understanding and appreciation of the biosciences;
3. To identify, analyze and solve problems relating to research, development and commercialization of biosciences;
4. To provide career support via the BayBio website;
5. To centralize life sciences information researches.

Built by the founders of biotechnology in the Bay Area, BayBio was founded in 1990 by consortium of universities, public officials, educators and bioscience executives to foster a regional climate "in which bioscience can continue to flourish". In its early years in Oakland, BayBio was a renowned centre for the bioscience job seeking community; the organization then moved its "information headquarters" to San Francisco in July 1999.

BayBio brings the bioscience community together into collective exchange information and ideas; BayBio also provides regional visibility opportunities. As the only bioscience organization in Northern California, BayBio serves the entire region's life science companies (nearly 820), a dozen of private research institutions, nine regional universities, and local government agencies.

Entrepreneurial Culture

Entrepreneurial culture means the scientist 'attitude to look at the scientific side of researches also at the commercial exploitation of their results, even if a significant part of the knowledge produced in universities and research centers can be codified *e.g.* patented or published in journals and broadly diffused and exploited, tacit knowledge, which is produced in research labs, but also "embedded" in learning processes and procedures must be equally considered. Starting a new venture, however, is a matter of entrepreneurship rather than science or technology, and requires also managerial competences.

To help the diffusion of the entrepreneurial culture among researchers, the following practices can be highlighted:

1. To teach managerial courses to science and technology graduates. Scientists often fail in the right evaluation of the commercial possibilities of the research result. They also lack the necessary managerial competences in order to define the industries the long term. Teaching managerial courses to life sciences graduates, even if with a low level of specialization, may make them "familiar" with managerial approach and may help them in understanding and better evaluating the external support by " professional" managers;
2. To provide MBA (Master in Business administration) and related courses dedicated to life sciences. The characteristics of business models of biotech industries are such to require "tailored" management solutions, that may be made available also for scientific;
3. To favor entrepreneurial experiences among researchers. This task may be performed with the possibility for academics to try the entrepreneurial adventure with the option, in case of failure, to go back to the previous academic position.
4. A competition for research grants. A competition for research grants may be introduced at national or local level, ranking universities and public research centers

by looking at their scientific outputs and also at to their attitude to the commercial exploitation of such outputs. This way to èvaluate the scientific activity, adding the perspective of commercial exploitation, reaches the twofold objective to reward leading-edge centers and to force them to improve their exploitation mechanisms.

Biocom currently offers two professional development courses: "Back to basics" and"From the laboratory to Leadership: Developing Scientific and Corporate Leaders". "Back to Basics" aims to educate biotech researchers and employees with basic concepts and techniques utilized in the field of biotechnology, combining lectures and with hands - on lab activities.

"From the laboratory to leadership: developing scientific and Corporate Leaders" is designed to train managers in scientific field with strong business leadership skills. Both lectures and team projects are used to teach communication skills, goal setting, time management and conflict management.

The Exist programme is a comprehensive initiative promoted by the German federal government in 1997 and still ongoing. The initiative is directed to high-tech start-ups with four main objectives:

1. The permanent establishment of a "culture of entrepreneurship" at universities;
2. The consistent translation of academic research finding into economic wealth creation;
3. The targeted encouragement of the great potential for business ideas and start-ups personalities at universities;
4. A marked rise in the number of innovative start-ups.

Among the actions carried out in the programme, particular relevance has been put on "entrepreneurial courses". These courses, with class sizes of nearly 20 post-graduate students with both economic and scientific backgrounds, concern the development of real start-up projects. The Fraunhofer Institute has scientifically supported the initiative, which focuses not only on biotechnology but also on IT and engineering sciences. Up till 2002, over 100 applications have been submitted and almost two-thirds of them have been approved for the actual establishment of new companies.

Mechanisms of Attract Key Scientific People

In order to develop efficient and effective mechanisms to attract key scientific people at universities and research centers. The following practices can be highlighted:

- To support research networks, research organizations and enterprises, in the provision of structured global schemes for the transnational training and mobility of researchers, and the development and transfer of competencies in research including those relating to research management and research ethics;
- To provide the means for research terms of recognized international status to link up, in the context of a well-defined collaborative research project; in order to formulate and implement a structured training programme for researchers in biotechnology. Networks aim to achieve a critical mass of qualified researchers and to contribute to overcoming institutional and disciplinary boundaries, notably through and promotion of multidisciplinary research;
- To support cooperative research projects among actors at international level, a major outcome of such kind of projects is enhancement of the competitive position of the actors involved.
- To support the "creation" of leading-edge scientific base. Enhancing the specialization of each university or research centre in a particular scientific or technological field, in which being able to reach and excellence, may, even if indirectly, be a strong mechanism to attract key scientific people worldwide. Scientists are attracted by the scientific "curiosity" and

by the possibility to carry out the research activity in a state –of- art environment.

As in the case of the diffusion of entrepreneurial culture, also concerning this issue, actions can be undertaken by the central government.

Measures of the European Commission to Stop Brain Drain

Based analysis of career prospects in the EU, the communication "Researchers in the European Research Area: one profession, multiple careers" identifies factors that impact on the development of careers in R&D, namely training, recruitment methods, employment conditions, evaluation mechanisms, and career advancement. The Communication proposes concrete steps to encourage and structure improved dialogue and information exchange with researchers and to establish a genuinely competitive research labor market at European level. In order to achieve the objective of raising Europe's investment in research to 3% of gross domestic product (GDP), as decided at the Barcelona European Council meeting in March 2002, the EU will need 700,000 extra researchers. There is therefore an urgent need to improve the image of researchers within society, attract more young people to scientific careers and foster researchers' mobility across Europe and back from other regions in the world. There are still some major obstacles to overcome, including in particular difficulties in cross-sector mobility such as moving from university to private business career, and the additional problems encountered by researchers attempting to embark on careers in universities outside their own countries.

Initiatives Set Out in the Communication Include

1. The launch of a "European Researcher's charter", for the career management of human resources in R&D;
2. A "Code of conduct for the recruitment of researchers" at European level;

3. The development of a framework for recording and recognizing the professional achievements of researchers throughout their careers, including the identification of tools aimed at increasing the transparency of qualifications and competencies acquired in different setting;
4. The development of a platform for the social dialogue of researchers;
5. The designing of appropriate instruments in order to take into account the necessary evolution of the content of research training;
6. The development of mechanisms to ensure that doctoral candidates have access to adequate funding and minimum social security benefits.

Industrial Driving Forces

Other than the scientific base, the industrial base is also an absolute requirement to build a geographical concentration of new six industries factors can be identified as influencing the birth and development of industry

- Presence of industrial base;
- Existence of success stories in biotech;
- Attraction of new sites of other industries.
- Integration among industrial actors;
- Support to R&D outsourcing processes and industrial spin-offs;
- Mechanisms to attract key managerial and commercial people.

The creation of an industrial base (given some exceptions) and the existence of success stories in biotech can be identified as context factors, representing a heritage of the past of the area.

The other factors are real actionable driving forces, and also in this case the objective of the related actors should be the creation of a "virtuous" circle.

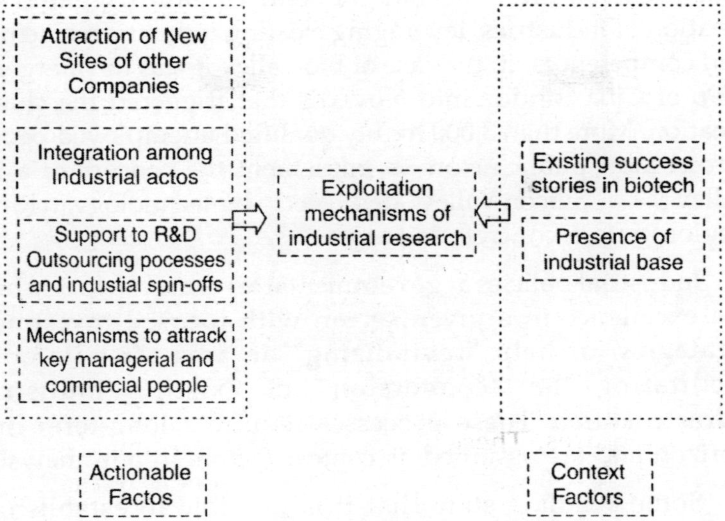

Fig. 9: Industrial Driving Forces

Attracting new sites and/or key managerial and commercial people allows the increase respectively of the industrial "mass" and of its quality In the area. The same results can be achieved by supporting the R&D outsourcing processes of large companies as well as the spin-off mechanisms. This stimulates the direct creation of new research-based companies as well as the one of service and supplier companies. Finally, benefits of co-localization can be actually exploited favoring the integration among industrial actors.

Presence of Industrial Base

For the development of biotech industry Presence of strong industrial base required. This is true for two reasons: (i) a strong industries base in the biotech sector represents a "dedicated" market of the research results of the universities and research centers as well as to small DBFs; and (ii) a strong industrial base represents a trigger for the creation of new industries both directly, through the mechanism of industrial spin-offs, and indirectly, favoring the establishment of suppliers and service industries as well as new core biotech industries. The restructuring of the operations at Pharmacia in Uppasala that occurred after the merger with Upjohn triggered the creation of industries, leveraging existing industrial structures and competences. In the case of Bio valley it was the merger in 1996 of Ciba Sandoz into Novartis that triggered the cluster creation. More than 3,000 highly qualified unemployed people forced local public actors to implement the concept of a Bio valley (*i.e.* a "Silicon Valley" dedicated a biotechnology in Rhine Region) into a concrete initiative.

Industrial policies at governmental level may help reaching the excellence in a given sector, with focused investment strategies, or help "revitalizing" an area after its crisis facilitating the "conversion" of existing industrial infrastructures. These processes require a long-term time horizon and are assumed as context factors in our analysis.

Some scholars state that it is possible to establish ex abrupto an industrial base through the direct "entrepreneurial" interventions can be found in the rapid development of Japan, in the past, and nowadays of China and India. In research driven sectors however, this still appears quite difficult to create. In the biotech sector, two particularly interesting cases of rapid development of the industrial base are Taiwan and Singapore. In both cases, however, governments decided to support initially the establishment of manufacturing centers, which mostly require less scientific competences and can be easily separated by the R&D centers, aiming at creating a favorable industrial background for the further development of local research activities. In Taiwan the support by the government consists in a huge funding activity, which accounted globally for US$5 billion during the period 2000-2005, and in dedicated industrial policies (*i.e.* tax incentives for new sites) one of the main objectives in the development of Taiwan's strategy is to establish the area as a link in the international community between R&D and the commercialization of products, acting as a global service and manufacturing centre for biotech. In Singapore, similarly, the evolution of the biotech context is strongly embedded with the activity of local government that has structured a dedicated central agency (EDB, Economic Development Board).

Supporting Driving Forces

Creating a constructive "environment" for the growth of the biotech company, the aspects analyzed here apprehension:

- The legal framework;
- The attractiveness of area;
- The presence of dedicated support infrastructures;
- The public acceptance of biotech activities;
- The international promotion

The positive occurrence of such kind of factors in the geographical area strongly contributes to trigger and to "spin" the "virtuous" circle analyzed with view to the scientific and industrial base and also facilitates and attracts investments, thus strengthen the whole scheme of driving forces.

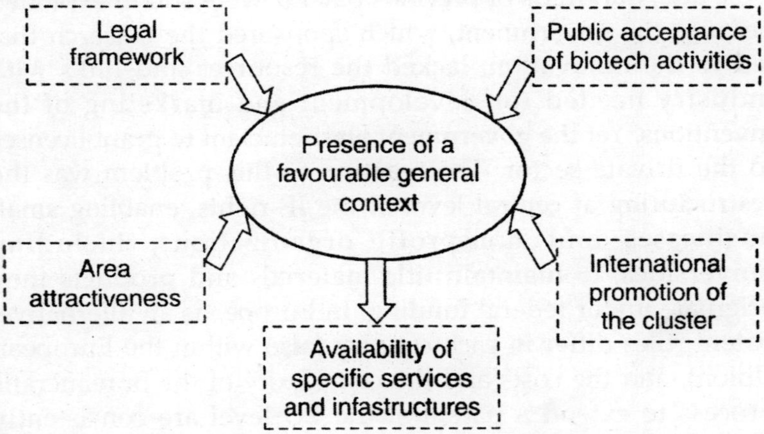

Fig. 10: Supporting Driving Forces

Legal Framework

A distinct and suitable legal context is unquestionably a requirement to the development of a science-based, highly innovative company.

Most important issues regarding predominantly the biotech company are: IP rights and bio labeling (*i.e.* the set of rules which regulates the labeling procedures, particularly for food products).

(i) IP Rights

In the case of academic researchers, the rules regarding the IP rights may correspond to a most important problem in the development of the research results. The definition of the legal context is first and foremost due to the central government; furthermore, each university may define internal rules, aiming at sustaining the development of research results. Therefore, the analysis of the legal framework should be conducted at two levels:

(i) The national (and/or regional) legal context; and

(ii) The local set of rules (*i.e.* the set of rules within universities and research centers.

With reference to the first level, major differences can be found in the US-EU comparison. In US, up to 1980 (Bayh-Dole act), hundreds of precious patents were still unexploited because the government, which sponsored the research that led to the innovation, lacked the resources and links with industry needed for development and marketing of the inventions. Yet the government was reluctant to grant licenses to the private sector. The reaction to this problem was the restructuring at central level of the IP rights, enabling small businesses and non-profit organizations, including universities, to maintain title materials and products they originate under federal funding. In Europe, as an alternative, patent rules differ in each country (also within the European Union), and the costs and the complexity of the bureaucratic process to extend a patent at the EU level are consistently higher than in US.

(ii) Bio-security and Bio-labeling

Bio-security and bio-labeling rules currently have an impact restricted to agbio and nutraceutical function and do not "interfere" with the development of pharmaceutical applications. However, it is deficient in of a clear legal definition, for example, of the use of GMOs, may have a depressive effect on the full exploitation of the scientific innovations related to biotechnology

Potential Support Measures are Following:

- To define IP rights of researchers and universities enabling inventors to retain and exploit commercially, through close relations with the industry, the outcomes (products and technologies) of their researchers;
- To speed up the related procedures especially for patent approval;
- To define, particularly in Europe, a common legal framework in order to exploit the advantages of integration among countries.

The US was the first nation to face the problem of IP rights, and already in 1980 precise indications came out both from the Congress and the Supreme Court.

Particularly two milestones put the bases for the future growth of the biotech sector.

(i) The Bayh-Dole Act, formerly the "patent and Trademark Act Amendments of 1980"; and

(ii) The diamond vs. Chakrabarty decision.

The Bayh-Dole Act created a uniform patent policy among the many federal agencies that fund research. It enabled small businesses and non-profit organizations, including universities, to retain title materials and products they invent under federal funding.

Congress perceived the need for reliable technology transfer mechanisms and for a uniform set of federal rules to make the process work. On major impetus for the bill was the lack of capability of the federal government to transfer technologies for which it has assumed ownership. The few federal agencies that could grant patent title to universities were over regulated, with conflicting licensing and patenting policies. Technology transfer under those conditions was operationally prohibitive for universities and made them reluctant to enter the technology arena. The stability provided by the Act has spurred universities to become involved in transfer of technology from their laboratories to the marketplace. The ability to retain titles and to license their inventions has been a healthy incentive for universities. Such incentive is needed, since participation in patent and licensing activities is time consuming for faculty and must be done in addition to research and teaching priorities. A 1997 survey conducted in US by AUTM (Association of University Technology Management) reports that 70% of the active licenses of responding institutions are in the life sciences, particularly in biotechnology. Most of the inventions involved were the result of federal funding (for example: the Cisplatin and Corboplating cancer therapeutics, Michigan State University; the Haemophilus B conjugate vaccine, University of Roachester, the Recombinant DNA technology, central to the biotechnology industry, Stanford University and University, of California).

Chapter 7

BIOTECHNOLOGY AND BIOETHICS

By now it is known fact that modern biotechnology has had a great impact on medicine and agriculture. It can only be predictable to have an even more dominating impact on future science and technology. Its impact is not inadequate to the technical impact that these advances have upon industry, medicine and agriculture any technology influences society and one can anticipate that life science technology potentially has the greatest impart.

Standard shifts include the switch to biodegradable products, industrial pressures to restructuring scientific in sequence sharing, the standard of sustainable and limited economic growth and the standard of interference in nature rather than surveillance and contribution in it. Biotechnology has also been a channel to the reflection of bioethical issues and the two words biotechnology and bioethics have co-evolved.

Criteria to Evaluate Whether Biotechnology Research is Ethical

In determining whether any given application of biotech is ethical there is some criterion, which may be useful. Even though it is promising to develop useful arithmetical scoring system as has been attempted for animal experiments. They are still only guides and may be more useful in directing attention to the better and more ethical design of experiments. There are only some criteria important to assess the ethnicity of an experiments or applications. There are some questions arises they are as,

What are the benefits? Is it life saving? What is an acceptable level of risk? Human benefits are greater than monetary benefits.

Humans should not harm. For human experimentation one should follow ethical codes. One should not cause pain to animals protect animal rights as much as possible & should develop non-animal alternative for research. With this environment should be protected.

Environmentally sustainable technology should be used. Pollution should be minimized.

Biodiversity should be protected. Endangered species also is protected.

Public and scientists should be informed and educated the all dimensions of the projects, scientific social, economic and ethical aspects of technology.

Bioethics is the study of ethical issues connected with life, including medical environmental ethics.

Bioethics

There are global, regional, national, community and individual issues, which are related to ethics. Ethical issues raised by biotechnology that involve the whole world and a single person.

A global crisis such as global warming may be aided by global applications of biotechnology for example to reduce net atmospheric carbon dioxide raised by dipping emission or increasing biomass however excess consumption and energy use can only be solved by individual action to reduce energy use. The introduction of new organism or of an unstable genetically modified organism (GMO) into the environment this is a regional issue. Other ethical issues arising from biotechnology that are attention of as individual issues such as genetic testing or use of gene therapy, also have public inference. Why we need ethics? Or what principles and factors are crucial for guiding decision making especially over such a diverse spectrum of issues? Are the question that may occurred in one's mind Medical ethics involves decision making on a individual level it apprehension the patient and the health care professional especially the physician. At a supplementary level missing may be many others who will be ultimately affected by such questions as the cost of very expensive treatment that takes funds away from other patients. At this level higher policy making is required, as in the case of issues such as environmental risk or intellectual property protection policy. All people are different as we look at our faces, sizes and the clothes that we chose to wear. Choices that we make are also different. In a democratic society we are acquainted with that we have a responsibility to allow people make their own choices. The challenges of new technologies and increasing knowledge, the challenge of respecting people as equal persons with their own set of values is a challenge for all. This is expressed in the language of rights, by recognizing the right of individuals to make choices.

Rights

Human rights are critical to maintaining human dignity. The acknowledgment of human rights has altered the circumstances in many countries. Many countries in the world have signed the UN declaration of Human rights. As we all have a right to be implicated in decision about our country the liberty of religion or speech, to raise a family, to share in the reimbursement arising from scientific advances. As human rights have always are difficult to define. The temperament of relationships between people in power and people without power with respect for personal rights should change from being characterized by dictatorship or paternalism to appropriate a partnership.

Law and Ethics are different. Ethics is a higher chase doing more than the law requires. The law is needed for protection of people and it sets a minimum standard. But one cannot resolve good moral activities by setting cases in a court of law. One should need to think of medical litigation or environmental damage penalties, which can lead to huge sums of money being paid for accidents (or negligence) which cannot really by compensated by monetary reimbursement. The elucidation is to have more careful and moral physicians companies, lawyers and politicians and the substitution of monetary balance sheets by ethical values as the primary cause of decision-making.

The chase of improved medicines and health is the most cited justification for this. It is better to attempt to do good than to try not to do harm. A disappointment to challenge to do well, working for people's best interests is taken to be a sin of exclusion.

The duty of Governments have to offer their citizens the opening to use new technology on condition that it does not violet other fundamental ethical principles. People should be offered the option of using new technology in medicine and agriculture. And such applications should make providing internationally accepted ethical and safety standards are applied.

Those who possess life-sharing technology, in medicine or agriculture have to share their technology with others who need it. This is applicable to biotechnology companies also may hold patent rights on particular processes they must contribute to it with others even if they cannot pay for it. This may mean that companies share developments with developing countries to give new drugs to individuals too poor to purchase them and many divergences with alleged business ethics.

The people who do harm even if the motive was to do well, the laws of society generally attempt to penalize. There needs to be a balance between the areas of science and technology where one can expect both benefits and risks. Risks must be balance with benefits. Technologies then apply their comparisons to their own behavior. In determining govt. policy do not harm is a very broad term, but is the basis for the philosophy of justice & privacy & humanity. It can also be expressed as respect for human life & truthfulness.

Biotechnology and genetic engineering are providing many benefits but there are also many risks. It is also unclear who will really benefit the most. All people of the world can benefit if it is used well, through medicines, and more environmentally sustainable agriculture. On the other hand Biotechnology inventions that allow developed countries to become self sufficient in many products will change the global trade balances and affluence of people. Because of political instability & war developing countries cannot export products. The make use of of enzymic conversion of cornstarch into high fructose corn syrup causes serious harm to the economies of sugar exporting nations and may by now have caused political unsteadiness and global issues.

Global food safety and environmental values should be swiftly developed to make certain that all people of the world contribute to their protection and no country becomes a testing ground for new applications.

Individual self-sufficiency comes above social attention one need to keep in mind that the reason for shielding society is because it involves many human lives, which must all be appreciated. Individual liberty is limited by respect for the self-sufficiency of all other individuals in society and the world. People's well-being should be promoted and their principles and choices appreciated but uniformly which places restrictions on the search of individuals self-sufficiency. One need to consider well being of future generations which places restrictions on this code globally, no single country should practice policies which hurt people of any country. This means they should have the freedom to decide major issues regarding their life and is behind the idea of human rights. Division of self-sufficiency is some independence to choose what to do as long as it does not harm others, also called individual liberty or privacy well being includes the code of do not harm to people and to work for people's best well-being.

The area of biotechnology patent policy should be examined in light of public view and the principle of justice globally. Naturally occurring genetic resources should be used common any one individual or company should not own it. For further scientific research some patent protection for specific applications involving biotechnology need to be protected to encourage further research and to make the results of such research immediately open.

Awareness of Ethical Biotechnology

Something, which is morally insulting to the majority of people in a country or region or worldwide, is judged to be dishonest and is likely to be banned. Immoral is frequently also unethical, despite the fact that unethical practices are repeatedly tolerated by a society and thus definition of moral would say that they are morally adequate because it is common morality. The generations who will have to live in a polluted and changed world one would say it is unethical. On the other hand this status are common morality to a preponderance of the people living in the rich countries even if the proportion may be falling and it is morally unacceptable to the poor or developing countries.

Both nationally and internationally decisions may be made democratically in a country if a agreement supports them if the rights of minority groups are valued and if it makes sense in the long term. On the other hand may be all decisions made in this way will not be ethical. Society can make unethical majority decisions and will continue to do so. In the area of biology and genetics one should not forget the unethical obligatory eugenics that swept the world in the first half of this century. More than 40 countries made laws to implement compulsory sterilization and discriminatory immigration policies.

Law is often based on the so called common morality of a country and in the area of biotechnology one can see varying laws established by different countries. Germany prohibits research as a criminal offense for the use of human embryo. British permits standard research when the embryo is 14 days old. The laws on the restricted use or discharge of genetically engineered microorganism vary between different countries due partly to different public awareness of risks.

Governments and companies occupied in Biotechnology research have become watchful in their monitoring of public view. In the case of governments they are not reelected and public resistance to companies can be expensive in terms of time delays and lost sales as most people receive information via the mass media especially the newspaper and Television. The media have a large accountability to converse science issues well and scientist should also update people about science. The media have a liability to present balanced information, on the benefits and risks of alternative technologies and do this autonomously of profitable interest. Groups who have a special interest such as political groups and other groups whose members spend time to publicize their opinion and who can get media coverage of their views can influence public opinion. The media may also have their own special interest.

Mixed Awareness of Benefits and Risk

Many of view polls have imperfect significance because they ask set questions with set response allowing little room for free response. The responses are often indicative and cannot give us the real picture of what the public is thinking. The additional time exhausted in study of free response question may be well worth the investment if the fundamental reasoning is to be determined. Even the use of questions looking at the balance of benefits and risk are more useful than asking single questions.

Privacy

The importance on privacy is very important. Personal information should be private. One must be cautious when using computer databanks that include personal information and if they cannot be kept confidential the information should not be entered into such a bank.

A characteristic of the ethical use of new genetics is the privacy of genetic information. This is one of the outstanding features of the accessible medical practice that needs to be toughened. It is not only because of esteem or people's self-sufficiency but it is also desirable to preserve trust with people. If one break a person's confidence then cannot be trusted. If medical insurance companies try to take only low risk clients by prescreening the applicants there should be the right to refuse such questions. The only way to ensure proper and just health care is to enforce this on employers and insurance companies or what is a better solution, a national health care system allowing all access to free and equal necessary medical treatment. One need to protect individuals from favoritism that may come in flawed would, one that does not hold justice as its peak.

Human Exploitation

There is also apprehension about human exploitation of these techniques, which could be eased by additional assurance over who uses these techniques. For human beings another major response we concerns about eugenics and cloning. This apprehension may be eased by the overture of laws. It possibly will be good to preserve a high level of such fears in society as the most efficient method to avoid future abuses of biotechnology from being made. On the other hand it is significant that people become skilled at to distinguish medical uses of genetics from cultural applications that are linked with world eugenics. They can already distinguish between application involving different organisms. In bio warfare research, Biotechnology is being functional to military research to develop biosensors to become aware of poisons and to perk up immunity. If the purpose is self-protective and the results are scattered to the general public, globally, one could find ethical explanation for such research. On the other hand such research would divergence with ethics if it is not shared with all people and if there is any possible escalaration of disgusting biological warfare research *e.g.* biotechnology will be applied to bio warfare development by individual terrorists. Additionally any research to improve immunity will be useful to face infectious diseases, and in the future as the increased incidence of UV light may decrease immunity. Biotechnology should not be misused

Animal Rights

The moral standing of animals and decisions about whether it is ethical for humans to use them mainly depends on several key such as domestic attributes of animals, the capability to think, the ability to be aware of family members, the ability to feel pain and the state of being alive. All will distinguish inflicting pain is bad, so in use of animals one should avoid pain. It is believe that we evolved from animals so one should think that some of the attributes that suppose humans have which present moral value on humans may also be present in some animals. Even though one cannot draw black and white lines can say that because some primates such as a whales and dolphins appear to possess similar brain features, similar family behavior and grief over the loss of family members to human they acquire higher moral standing than animals that do not show signs of these. So if one can attain the same end by using animals that are more primitive than these such as other mammals, then we should use the animals at the lowest evolutionary level appropriate for such an experiments or for food production. Use animal cells rather than whole animals or use plants or microorganisms for experiments or for testing the safety of food.

Environmental Ethics

Humans have relations with the environment and in fact depend upon of the environment for life. There are varieties of product such as food, clothing, housing, fuel and medicine, are beneficial to human. The variety of uses also supports the protection of the diversity of living organism, biodiversity. The ecosystem is precisely balanced and the introduction of new organism into the environment may disturb that balance. This is a further concern raised by genetic engineering. It is that nature has worth in itself because it is there. Species should not damage unless it is absolutely necessary for the survival of human beings (not the luxury of human life). Before one can disgustingly change the characters of individual organisms or make irreversible changes to the eco-system and human society one need to know what are these supposed limits of changing nature, there will be future possible applications of technology, which are against common morality. A reading on the metaphors discovered in open questions about life and nature in different countries of an international bioethics survey showed universality of thinking about these concepts. At the lowest end of the scale of ethical status microorganisms are generally placed because the only domestic character they have is a human esthetic perspective means that the only disagreement usually applied to them is human utility.Human beings are challenged to make ethical decisions and to stability the reimbursement and risks of alternatives they have to. The benefits are great, but there are many possible risks. Any effort to develop global bioethical approaches must engage contemplation of the values of all people. Widespread ethics may be possible to a noteworthy degree, however one can even have difficulty in universal recognition of basic laws such as those respecting human rights. The survival of international environmental laws *e.g.* the law of the sea and charters of human rights is some support for the future programs of limited universalism. Cross-cultural bioethics involves shared perceptive of a choice of cultural, religious, political and individual views that people have. The variety of individual perspective, in any one culture appears to go beyond, the difference between any two *e.g.* in any culture one can find people zealously apposed to induced abortion and these who support it as a right for a women's preference. The opinions expressed in the responses to an international Bioethics survey conducted in Australia, Hong Kong, India,

Israel, Japan, and New Zealand. The Philippines, Russia, Singapore and Thailand suggest that people in these diverse countries have a similar variety of reasoning. This type of research provides to better understand the reasoning of people. People in many countries do allocate the same hopes and horror, which strengthens the call for international principles.

Ancient Times and Current Bioethical Conflicts in Biotechnology

There were assumption that genetic manipulation was interfering with nature, or that it was curse to God and also genetic manipulation of humans and animals as unethical. In the US survey 46% said that we have no business interfering with nature, while 25% disagreed (OTA 1987). Even though many scientist respond to people with these views as ridiculous but it is remarkable that about 16% of the scientists and teachers who found these techniques unacceptable also shared these views and these reasons were also cited regarding genetic manipulation of microbes.

In Japan 12-16% of the public who were worried about consuming products made from GMOs said that such food stuffs or medicines would be aberrant, while in New Zealand the values were much higher. These results were confirmed in 1993 international Bioethics survey. Australians were similar to New Zealanders with other Asians expressing this idea less like Japanese. While generaliy one can say such foods are just as natural as foods made from any modern crop or animal breed 10-12% of scientists also said this. In a 1980 public opinion survey in Britain 70% agreed that natural vitamins are better for us than laboratory made ones. While only 18% disagreed. Supplementary studies of attitudes to consuming products of genetic engineering exposed similar fears in the Netherlands. The use of chemical pesticides and preservatives during crop growth food storage and processing, should be avoided

There is a changing concept and varies between individuals, religions and cultures about nature. Societies becoming urbanized they lose touch with nature. However there is also a modern trend to buy products from organic farming or preferences for free-range eggs over eggs from battery-farmed chickens. All people have some boundary in the degree to which they hold up changing nature or the purpose of technology. Within the intervening time scientists as well as the public distinguish restrictions to what is good enough or ethical, biotechnology, and supplementary research is needed to verify what these restrictions are. Biotechnology may essentially be competent to make these areas more natural by reducing the use of chemicals in agriculture, food processing and medicine

Ecological and Environmental Risks

There is mounting apprehension about ecological and environmental issues. The phrase sustainable technology has been added as an ethical criterion of technology. With the distorted U.V. light and predicted greater than before temp and climate change, the existing circumstances are not sustainable; to a certain extent one require to look at the steady state that may be possible to steady in an additional century.

On the other hand the previous action was less different the future world will be and sustainability has become recognized as an ethical principle of technology. Any biotechnology, which is in conflict with this principle, will be called unethical Biotechnology has prospective for ecological enhancement if applications are embattled with that aim. The ecological apprehension that many people have, refers to the introduction of new organisms into the environment, including GMOs, ecological studies are desirable and monitoring of release of GMOs to increase data to allow calculation of the ecological impact of such introductions. Some data are by now accessible, after over 1,000 fields tests of GMOs already performed, but awaiting protection has been established for each type and species of GMO in question one should not have commercial introduction on a large scale.

Future Bioethical Conflicts in Biotechnology

Some diseases are natural that it may not involve human action but it can be apparent an aberrant when it is curable. In medicine there are ordinary treatment and extraordinary treatment. Biotechnology may not have an elementary variance with nature and many would see new technology as a souvenir from God, but one must not mystify us with God. This means that there should be limit for the introduction of genetically engineered microorganism into the wider environment despite the fact that many may arise via the discriminatory pressures present in polluted environments. On the other hand, abolition of disease is ethical biotechnology; there should be no objection to the abolition of human diseases such as small pox. There may be more contests about the abolition of recessive disease-causing alleles that may have some unknown benefit like the sickle cell anemia trait, which offers better defense against malaria. But no baby should be deprived of the possibility of therapy for the disease, be it by safe gene therapy or other options. Though carriers may want to abolish their risk of transmitting a harmful allele to their offspring- what is called germ-line gene therapy. This is another future conflict that faces are yet if some countries endorse policy that make illegal germ-line gene therapy for the present.

Chapter 8

PRODUCT GROWTH IN BIOTECH

Structural Risk Evaluation of rDNA Products and Consumer Approval of Products

Since the Asilomar conference in 1975 Biotechnological products base on rDNA have been a subject of dispute at many levels. Debate was among assiduous scientist who required thinking about the cost of new techniques in molecular biology before applying these in the edifice of new combinations of DNA. Previously during the Asilomar conference this debate became steamed up by a number of unempirical arguments and personal interests. In the media some journalist started to portray this technology as a modern adaptation of Cerberus or the molecular adaptation of the monster of Frankenstein. Regrettably the scientific community was not very well prepared for this debate and did not react homogeneously. Furthermore, the scientific community started to evaluate the risk related to rDNA technology to other risks *e.g.* the risk of impulsive or calculated mutations or conventional breeding consequential in hazardous combinations of DNA. Comparison of such types was scientifically inaccurate and led to auxiliary perplexity among the public. A large amount regrettably compared rDNA technology with nuclear technology this has been extremely misleading. As far as nuclear energy is concerned it is the intrinsic hazard and it effects may be very large and will affect all living species. But for rDNA technology the intrinsic hazard will depend on the nature of the foreign gene, the host cell and various other factors and it will range from zero to one. All through the preliminary discussion it was understood that in general the hazard of rDNA products was not zero and so the debate determined on the physical and biological suppression of the experiments to reduce the risk.

Presently following the Asilomar conference in most countries board were created to set up rules to make certain that the intrinsic hazard of the foreign genetic property was low and that biological and physical suppression would assurance that the probability that the hazard would become evident outside the laboratory would be enormously small. Individual manufactures made the arguments that the rules proposed to evaluate hazards and risks related to rDNA could put at risk their future were not correct and contributed to a further divergence of views. This is why there is a delay of the acceptance even in a rejection of rDNA technology outside the area of health care by the public.

The application of rDNA technology helps us to understand the basic biological processes. After 20 years Paul Berg (1974) and others concluded that the Asilomar conference were absolutely right in that rDNA technology would have an massive impact on mankind and in their appeal for a careful assessment of this technology since so little was known about it. A large number of discoveries that have been made since 1975 such as the existence of introns, the role of transpons in many species, RNA as enzyme the reprogramming of transcriptional competence, translation processes, protein folding and processes resulting in horizontal fluxes of genes DNA between species. Even though at present still much more has to be erudite the results of millions of rDNA experiments show that there is harmony in biology and consequently rDNA technology is as such not intrinsically hazardous. In addition there is at present ample perceptive of basic processes in biology to allow a balanced approach of the assessment of hazards and risks related to rDNA technology. Knowledge of molecular biology together with a structural approach to assess the diverse aspects of rDNA technology unconnectedly can give to end the uncertainty about real or apparent hazards and risks related to rDNA technology

An Assortment of Stages in the Development of an rDNA Process or Product

In a number of countries after the Asilomer conference established guidelines to evaluate the safety of rDNA experiments in the labs. They focused their efforts on creating frameworks for the categorized of rDNA experiments on the basis of intrinsic hazard of the foreign gene, nature of the vector used in the conversion procedures, apparent hazard base on the origin of the foreign gene, probable errors in the cloning procedures and the intrinsic hazard of the host cell.

Originally there were many differences in the views of this committee but after number of years almost all Western countries and Japan applied relatively the same rules for the assessment of the safety of lab experiments. At the end of the seventies the first guiding principle for experiments greater than the 10 litre scales were customary and these rules were fairly comparable in different countries. Pilot-plant and large-scale production rules were formulated in early 1980. For each suppression level general approval for the used strain, equipment and processing are necessary in most countries. An additional approval procedure must be started is quite arbitrary for the volume of 10 lit. Even though frequently the consent of direct and creation amenities and the process has been quite time consuming with a few exceptions The procedures for approval of new products, including rDNA products for pharmaceutical products are quite clear. rDNA production in food processing or for their use in food or detergent products or in other consumer products, nuclear procedures vary from country to country. In Western Europe (EEC), USA and Japan the lack of obvious and consistent guideline has clearly contributed to a considered delay in the introduction of rDNA products and resulted in certain companies reducing or even losing their interest rDNA

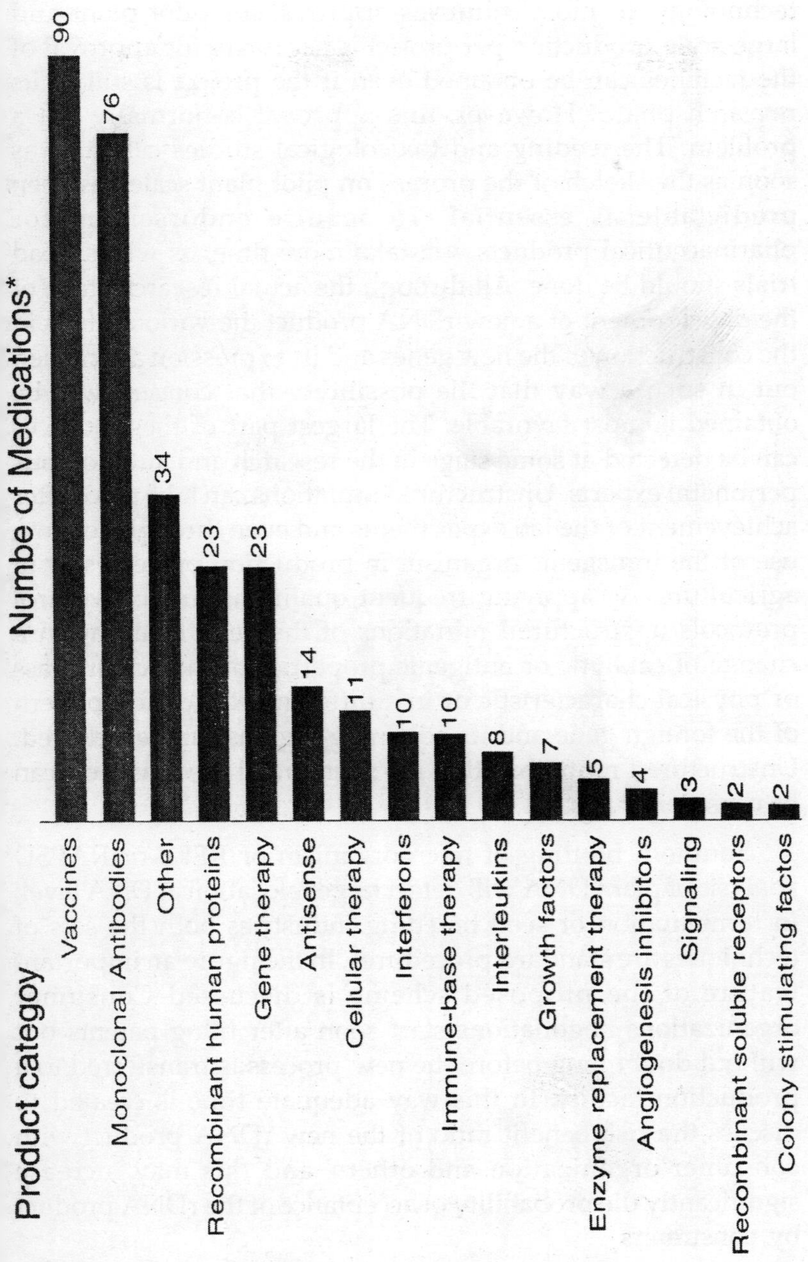

Fig. 11: Some Medications are Listed in more than One Category

technology. In most countries approval for pilot plant and large-scale production per project is necessary for approval of the facilities can be obtained even if the project is still in its research phase. However, this approval is normally not a problem. The feeding and toxicological studies can start as soon as the sketch of the process on pilot plant scale has been predictable as essential. To acquire endorsement for pharmaceutical products will take more time, as widespread trials should be done. All through the actual research stage of the development of a new rDNA product the various steps in the construction of the new genes and its expression are carried out in such a way that the possibility that consent will be obtained is most favorable. The largest part of these sources can be detected at some stage in the research and autonomous peripheral experts. Unstructured mutations can take place after achievement of the lab experiments and even through genuine use of the transgenic organism in production processes or in agriculture. By applying frequent quality assurance/ control protocols unstructured mutations of the gene resulting in a change of catalytic or antigenic properties or molecular mass or physical characteristic or in a different. Restriction pattern of the foreign gene and its adjoining regions can be detected. Unstructured mutations that are silent in all these aspects can be considered as safe.

Southern blotting of microorganism or RFLP or RAPSD analysis of plant DNA will detect large relocation at DNA level. In fermentation or seed breeding industries both the sets of techniques are standard procedures. In the figure an important feature of the proposed scheme is discussed Consumer organizations negotiations start soon after filing patents but without doubt long before the new process is transferred to a production facility. In this way adequate time is created to discuss the risk/benefit ratio of the new rDNA product with consumer organization and others, and this may increase significantly the probability of acceptance of the rDNA product by consumers.

Universal and Viable Aspects of Biomedicine Product Development

For biological pharmaceutical product development might emerge as a simple prolongation of a research project after an adequately valuable compound has been recognized. Persistent research turns into conventional testing of quality, safety competence and into analytical and process development. Development of a pharmaceutical product transfers on lively principle or compound from the reserve inadequate to research lab into a new environment to the stage of development. The compound is subject to a array of analytical, pharmacological, toxicological and clinical tests and trials in a multi-disciplinary approach. Then all methods, results, records of the materials produced are under severe analysis exerted by the developing institute itself and finally by several authorities.

Research & product development are habitually separated in different units in bigger organization as there are elementary differences between inventive researches and methodical, goal oriented development and also development requires particular knowledge's and experience. Most companies and organizations in the field of biological, pharmaceuticals are comparatively small in view of the fact that they traditionally function on a national scale. They do not have several products under development, which are subject to a similar development scheme. The methods are supplementary, composite and the materials created in the research phase are compactly passed on for development. In addition biotechnological products are still in their early years and pharmaceuticals from biotechnological processes are not yet well developed. Consequently there are many reasons why researches in these areas in industrial and non-industrial research organizations contribute to or perform product development. These scientists have to to know more than just their science to comprehend their new assignment

Universal Aspects of Bio-medical Product

Impractical opportunity concerning the effortlessness and interval of development may lead to unsuitable management decisions, which may even jeopardize the expectations of research oriented enterprises. Biotechnology success ventures do not only depend on superior research but also on a pragmatic evaluation of the probable products and on the aptitude to assume the required skills to successfully develop research results into commercial products.

Commercial Probability and Risks of Pharmaceutical Development

Pharmaceutical development is expensive and the risk of project failure is high during the process. The subsequent details and statistics may demonstrate the costs of and risk that can be anticipated. Even though these statistics do not discriminate between research and development and are not precise for biological pharmaceuticals (they represent mainly chemical entities for human drugs), the general picture is similar for all types of pharmaceuticals, and the implications for product development are essentially the same.

The Average capitalized cost to develop a human pharmaceutical product in USA amount to (US $231 million MASI *et al.*, 1991). Comparable figures for 1982 are US $ 91 million (LANGLE AND OLLELLI, 1983) and for the time between 1963 and 1975 US $54 million (HANSEN, 1979), on other hand, these significant increases in cost, which tended to double within ten years, imitate the mounting difficulties to find new and commercial attractive drugs, since these figure include the cost of aborted projects, on the other hand, or substantial part of the rising cost is due to the increasing dictatorial endorsement.

The standard time to develop a newly revealed compound increased continuously over the pass decade from less than 4 years in 1960 to more than 10 years in 1989 (KARIA *et al.*, 1992). As a corollary, the standard affective patient protection for human and animal health products approved in the USA between 1984 and 1989 was only 10 years and 7 months, including the possibility of reinstatement of patent terms of medical products of up to 5 years (VANGELOS, 1991). Longer development times mechanically diminish the efficient patent protection period and thus appreciably manipulate the earlier opportunity.

The number of new products launched and their sales expectations remained fairly constant despite the sharp increase in development cost. (LUMLEY and WALKER, 1992). It has been designed that in the UK even the best selling 10% new drugs involve longer than the efficient life span of their patents to recover the research and development cost; the majority of new pharmaceuticals cannot recap these cost even after more than 20 years (PRENITS *et al.*, 1988). Comparable figure with reasonable validity for animal health products are not available. Veterinary product development cost is absolutely lower, but the research revenues are also much lower, due to increasing registration demands, influenced cost follow the same trend as human health products.

Since 1980 the animal health market has stagnated. The global competition has increased, and the more measures of biological and vaccines try to internationalize their products and sales in order to out stability the price erosions by growth in sales volume. While vaccine procedures in the USA have been rather successful to penetrate European markets. European vaccines were prohibited from access to the profitable US market due to concerns about probable adventitious agents in these vaccines (*e.g.* FMD, BSE). Simultaneously European veterinary vaccines have to meet the terms with increasing dictatorial demands and the introduction of GMP (Good Manufacturing Practice) standards.

In the animal health care market relatively few really new and attractive products have been introduced. In animal health care the newly developed new and expensive human therapeutics do not play a major role. Prophylactic vaccines against the major endemic diseases are already available and are difficult to advance for a competitive price.

Assured high revenues cannot be attributed to the continuing high interest and investment in pharmaceutical research.

According to the grapevine the universal awareness of high profits is strongly predisposed by the commercial success of a few runaway success products such as the H2 receptor antagonist ivermectin or the recently launched erythropoietin. These are the exception whereas particularly expectant candidates, rendering very moderate success, appear to be more the rule.

The pharmaceutical industry is much more research oriented than other industries and has to deal with higher risks. Due to the high development cost and long development times, risks must be eliminated as early and accordingly potential. Before the expensive industrial development commences in the research phase risk elimination must be started. As high-risk research projects require much time, they must be reserved small and have to focus only on the main risk factors. Research projects should be clearly separated the more rigid and much more expensive development.

Just the once an attractive product has been acknowledged, the time factor becomes particularly important, and risk and cost are no longer the only appropriate criteria. Hence pace and effectiveness is of vital significance. An inconsequential development miscalculation that results in a delay of an assignment on the significant path for only one month can delay the entire project by one month. Roughly unavoidably it also adds 1/12 of the annual budget to the development cost. These in a project with an average annual budget of $2.5 million, a delay like this can cost more than $20, 00000. A part from these direct cost later accesses to the market can cause much higher losses.

Resourceful Market Study and Other Precious Milieu Information

According to a study management of new products by Booz, Allen and Hamilton conducted in 1968 standard about

70-90% of the money used up in research and development of new products is used for products that are failure. Industrial organization would not commit itself to an applied research or development project without investigating the general and market situation to reduce this failure rate. There is no reason why publicity funded research projects should not do the same. Project to start without assessment means being more concerned about doing the project right than doing the right project.

In a very slender logic researchers have a propensity to view market research. They do not distinguish the prospective of the approach thus missing pioneering ideas. Market research may in reality bring search ideas and help to keep away from most important mistakes also can help to identify interesting options. Thus market research adds information and uses it for a better appraisal of chances, risk and options. It should not be used to decrease the project idea to trouble-free information. The effectiveness of market research depends a good transaction on thorough interface between research and marketing and information replace in both directions.

For a basic appraisal of important market and other environmental factors does not need marketing experts. But if their advice is required, it should be by an interactive process not by presenting the idea for decision. The marketing expert will perhaps recognize the weak spots of the researchers offer during this process. The researchers with rational background information be acquainted with and appreciate the prejudiced assumptions of the marketing experts' appraisal.

The supplementary information usually requires changes of the direction or the priorities of the project; a researcher who is convinced of an idea is unlikely to consider such changes. Consequently researchers supposed to do their own market appraisal to develop estimation about the chances, risk and best applications of their approach. On the threshold it must be kept in mind that it is very easy to characterize an ultimate product, but this ultimate product is maybe the most difficult one to develop or may be an impractical development objective. The most comprehensive the collected information is the better it serves its principle.

Inventive market research should originally collect a few basic figures on the occurrence of the diseases or condition to be treated. Next step would be collection of information about current methods to diagnose and treat the disease or condition. It includes the products or compounds used, the frequency of their application, cost of the products used and the overall cost of the treatment. This information can be obtained by a specific literature search in the medical field or from physicians and clinicians. One can ask to clinicians for deficiencies in the current products or treatment schemes; such as side reactions lack of competence and for other medicinal treatments that have an impact on pharmaceutical products in this area.

An effortless appraisal of the likely cost of the new or improved product, compared to the reimbursement that it provides can be very revealing; still the figures are only rough estimates. Satisfactory cost for the new product may be deduced from prices of equivalent product classes or from unambiguous products that are presently used to treat the disease in question. These prices are likely to a give a rational approximation. A premium can be added to the basic figure depending upon the estimated value of additional benefits of the new product. This sort of simple calculation and estimate is particularly reasonable for veterinary products and for products that will contend with already accessible and recognized products. If the products or prophylactic treatment schemes cost only a few dollars the active ingredient is only worth a few cents. To get a rough idea about the upper limit for the cost of the active component in these cases, the market price of the existing product may be split up by the following thumb rule of calculation about 40-50% for cost of sales and a profit margin (before tax) 20-30% for quality control. Packaging and labeling and 10% for formulation for lower priced products in a competitive environment the active ingredient may account for only about 10-50% of the market price.

It is supposed that the planned new product can contend in terms of cost of manufacturing notwithstanding a more challenging process, or that the product offers considerable advantages which justify a higher price, it will be essential to present support for a high yield system or to focus on a evidence of the claimed compensation.

Market assessment for a product to be developed identifies the potential strength and weaknesses of a new product as well as those of competitive programs or therapeutic schemes. This assessment should include the patent situation and the competitiveness of the organization in terms of the skills and resources to develop manufacture and sell the product when assessing the chances and risks of the intended development product.

Points to think in planned and market oriented planning of research and development for biological medicinal product.

Universal Aspects

a) Disease occurrence, geographic allocation and occurrence

b) Current diagnostic and therapeutic measures.

c) Deficiencies of current treatment: efficiency, side effects, cost, practicability.

d) Prospective development of existing methods

e) Approval of prospective development: safety, cost, practicability

f) Interface with health care and suppression schemes

Competence Aspects

a) Prospective indications, scientific viability and risks degree of competence essential.

b) Systems and models to display competence

Safety Aspects

a) Potentially injurious effects of the planned products safety requisite: Similar or better than existing products.

b) Systems and models to evaluate safety

Viable Aspects

a) Product prices and cost of current treatment occurrence of use of existing product, likely cost of the proposed new product and treatment.

b) Cost/ benefits relationships:

Patent Aspect

a) Accessible patents and applications in the field chances for an own patent applications.

b) Availability and costs of patent support.

c) Cost/ benefit of patenting

Project Related Aspects

a) Other research and development projects in this area: Scientific competitiveness of the own project.

b) Specific skills and methods required Collaborators for specific problems to extend the scope of the project

Product Outline

A product outline should be recognized for each development product. It describes essential product qualities such as the indications, form of appearance, future application and use pattern competence, safety, and other significant aspects *e.g.* patent restrictions and the cost limitations for the compound or for the formulated product. The expected dates of product launch and sales expectations as well as the expected development cost are important elements of the product file. The cost-benefit calculation for the same development product may vary considerably between different organizations (HODDER & RIGGS, 1991) finds the calculation methods may be different.

Previously offered manufacturing amenities devoid of require for a key investment can appreciably manipulate the result of a cost-benefit study. In consequence a cost-benefit analysis in a circumstances and the assumptions that farmed the basis of the described situations should be precise and carefully taken into account, before general conclusion are drawn. In addition the aspect of providing information to assist in administrative, a particular product outline has more than a few other benefits. It creates a widespread vision about the path of the development by defining the aim unmistakably to all people involved, thus supplementary in stretched project planning without the necessitate for endless negotiations about the goals.

The procedures to resolve problems can be integrated in the development plan at the best, most suitable and cost effective time are critical aspects of the project. The product outline helps to identify changes in critical parameters (*e.g.* in efficacy) and interval from the decided development rate before, so that counter-measures can be initiated prior to these become too-costly, the soon after changes are initiated. That is it describe an ideal or almost ideal product It requires input from research, preclinical and clinical development, marketing and sales departments and in developed by an iterative, stepwise process.

Registration Necessities

The quality, safety efficiency and formal requirements for market are registration, requirement for medicinal products. For the period of last few decades pharmaceutical guideline in western world have become so multifaceted and specific that the task of observing them cannot simply be delegated to registration specialist. They are consulted all through the planning phase of a expansion of project and then they give advice, which laws, directives and strategy apply in a specific case and to what degree strategy must be adhered. The registration of medicinal product is governed by comparative values that cannot be clearly defined or limited for example, by ethical considerations to make certain the best possible quality and safety of pharmaceuticals.. Registration necessities are flexible and change with time, what is potential and practicable seen becomes a standard by which subsequent products will be measured. Which leads to continue and desirable improvements and requires continuous attention of those who develop and register projects. The most successful and beneficial pharmaceutical products, such as the virus vaccine, live poliovirus vaccine and other live vaccine, would probably not be remittable for a wide spread use in humans according to today's standard

Market endorsement or registration of a pharmaceutical product depends on the accomplishment of three main necessities

a. Quality
b. Safety
c. Efficiency.

The definitions for quality, safety and efficacy, diverge between product groups, centuries, authorities and between people within authorities.

Quality

The quality can drastically effected the safety and efficiency and advantageous product may be adequate, even if it has severe side effect keeping these independencies in mind, product development must deal with and highlight the three basic registration requirements in the same order. Although project risks are suggest having different strategies that concentrates more on efficiency. Without a defined quality, all safety and efficiency tests may be inappropriate and devoid of an ample safety, efficiency trials under natural conditions cannot be approved.

Quality is the test of high variability in particular of biological products and all variables of the manufacturing are evaluated. Product quality commences with the starting materials, and their attribute run through the entire manufacturing process and extends to the manufacturing facility and its infrastructure. The whole lot that can influence the product quality must be assessed tested and acknowledged. *e.g.* if one intends to produce a product constantly from a culture it must be specified what constantly means. If it is proposed to harvest a hundred times over a hundred days one has to prove that the product is still the same after the hundredth harvest and almost certainly even away from that limit to add some safety periphery.

It is essential to authenticate all the in process control and product quality control tests. All claims on product safety and efficiency depend on the particular quality if the quality is not reliable, safety and efficiency may not be the same. The data for product safety and efficiency are only lawful, if they were established with the material of the precise quality. The practical implications are that test material must meet the significant stipulation of data can be generated. The test for biological product has to be made by a process, which must not diverge from the final manufacturing process, is essential elements. If the manufacturing process is changed during the development all previous data with earlier material may be worthless, unless bioequivalence can be proven.

Safety

A product safety is an aspect of its production uses the disposal. Which includes the manufacturing location and their environments starting materials the active constituent of the product as well as metabolites, impurities and preservatives. The safety of manufacturing process and the facilities is a variety of regulations concerning the layout of facilities, work organizations, procedures and related controls and inspections. Wide-ranging national set of laws on work safety and biohazards, control measures, specific policy for manufacturing pharmaceutical products and quasi-legal GMP necessities have to be applied

A range of preclinical and clinical assessment that depends on the product and its use pattern addresses safety of the product itself for the target organism, the user or the environment. Local and efficient tolerance, acute and chronic toxicity, immunotoxcity and for verternity medical product also the ecotoxicity includes the series of safety to be assessed. Mutagenicity, tumoriginicity and reproductive toxicity testing may also be required for certain products or indications. It is necessary to know the action and fate of the product and its components within the body, after it has been administered for safety assessment of a pharmaceutical product in the target organism. Pharmacodynamic and pharmacokinetic studies,shows the results of influence of the choice and setup of safety tests to be conducted. Pharmacodynamic studies explore the influence of pharmaceutical substances on the human or animal organism such as their mode of action and mechanism of side effects along with dose effect relationships pharmacokinetic studies deal with the influence of the organism on absorption, distribution, metabolism and elimination (ADME) of pharmaceutical products in tolerable models and in the target species. There are side effects and can leave residues in animal that end up in human food. Their selection for a development product must be considered cautiously one uses ingredients with a well-established quality and safety outline or a variety of analytical and safety data have to be recognized for those excipients. Safety studies for new excipients may cost more than the safety evaluation of the biological product itself.

It is crucial to exhibit that pharmaceuticals for food producing animals do not have harmful effects on humans, particularly if residues of the active components or of excipients and of their metabolites may occur in meat, milk or eggs. This means that widespread studies regarding the safety for humans and residue exhaustion studies must performed with these products.

Products used only once or few times have to undergo fewer safety tests than products that will administer regularly or over a longer time. Less if any side reactions are tolerated for products proposed to be used in healthy individuals *e.g.* vaccines than for products that are used in diseased individuals. In any case reimbursement must clearly outreach any potential side effects.

Efficiency

Clinical trials in human are usually the most dangerous, most expensive and most time consuming developmental step and can account for more than one half of the development cost. There is repeatedly the misapprehension that controlled experiments in the target species, which shows that the product is successful, are sufficient to register a product Animal health products also have to undergo clinical or field trials to demonstrate their efficiency and safety under realistic conditions, these trials make known several previously unrecognized weaknesses of the product. On the other hand, clinical trials for veterinary medicinal products are much less vital, less expensive and usually less time consuming than human clinical trials. Due to the preclinical pharmacological, efficiency and safety tests performed in the target species, the clinical trials are mainly designed to authenticate preclinical efficiency data and to provide a broader basis for the safety appraisal. To find out efficiency of product clinical trials are an important base of claims. As reasonable consequences all indications for which the product is suggested all proposed routs of applications, treatment schemes, doses, relevant age groups and in the case of veterinary products all suggested species must be tested. In support of an efficient development it is often better to originally concentrate on only on or a few of the prospective indications to reduce the risk and to get the product on the market earlier, than to follow up all variants at the same time.

Technical Aspects of Product Development

Medicinal products must have a consistent quality in order to be dependably efficient and secure. Achieving stability in all aspects is an unwieldy technical and analytical task that takes much time. Scientist who ignores the technical aspects led to a consistent product quality.As there are increasing registration requirements for the products and the production process as well as environmental issues that add a noteworthy technical and analytical aspect to the development, this applies to products derived from biotechnology. The final product is generally much better defined and its manufacturing process can be better controlled but it requires the application of a variety of quite complicated techniques to do so. Therefore for new biotechnological products, processes new analytical methods have to be recognized. Which is a major and costly part of product development. Newly founded biotechnology firm's significant technical experience and practical knowledge about the application of authoritarian necessities is frequently missing and it must be developed. Near the beginning information and contemplation of these aspects may significantly shorten learning process.

Process Growth and Manufacturing

Process growth represents an association between research and manufacturing and adapts the research methods to the needs of production or develops new methods where essential.

Process developers must take several aspects into considerations. Their job would be much easier if researchers were conscious of these aspects.

I. Starting Materials and Excipients

Appropriate quantities with proper specifications and consistency, regular supply of starting material and excipients.

II. Safety and Environmental Aspects

Work safety, *e.g.* organic solvents, aerosols containing microorganisms. Waste material refinement and dumping.

III. Legal and Regulatory Aspects

Registration requirements for product and process licensing of facilities Good Manufacturing Practice (GMP) Infringement of existing patents.

IV. Economical Aspects:

a. Cost of goods
b. Yields
c. Recovery rates
d. Investment into equipment/ facility.

The active ingredient must be found and established. Regularly stocks like cells, viruses, bacteria must be recently generated and tested since those from investigate are not appropriate because of their quality or unclear history and documentation

Deplorable starting materials used in research must be replaced by those which meet the authoritarian quality and safety criteria indication to standard volumes of pharmaceutical ingredients and excipients should be made to decide whether a certain ingredient can be used. Standards are described in the pharmacopoeias of related countries in The Extra Pharmacopoeia of (Martindale 1993) and in the Handbook of Pharmaceutical Excipients (American Pharmaceutical Association/The Pharmaceutical Society of Great Britain 1986), The German Lexiken der Hilfsstofee (Fielder 1989) provides to large extent valuable information about their applications and safety. Mechanism used and standard for human food may also be considered as probable ingredients for certain applications. During the process dangerous components cannot be used at all or have to be removed. To avoid the presence of adventitious agents in the final product special conditions should be apply to the use of starting material of human or animal origin also process developers should make sure existing patents and patent applications in order to keep away from patent encroachment with any use or process patent. Regular attention is needed for economic aspects such as yields from cell culture or fermentation; product recovery after refinement and cost of materials and equipments Production cost should be calculate and updated on a regular basis. When a process is calculated an array of product safety, work safety, environmental safety aspects and other legal aspects must be taken into contemplation. GMP authorization is subject to assessment and requires invariable consideration and updating extensive documentation must be provided to act in accordance with these necessities.

Manufacturing amenities are accepted for their precise function, if they planned to manufacture another product with the same amenities then they must also be accepted and requires very severe measures to avoid any cross contamination and jumble. Whichever the amenities have to be separated or production runs for different product have to be performed at unusual times with concentrated cleaning and decontamination in between the ideal option is of course to develop a manufacturing process for a previously existing manufacturing plant. The new plant and the manufacturing process must be planned; built, equipped and approved first.

To develop a restricted process, which reproducibly gives a product with ample eminence, may take months or years, so it should be start in early in development phase. It also makes sure that no time is wasted by testing products of insufficient quality which may cause the data obtained useless for registration purpose. To change significant product specifications, in some cases it may be inescapable. If major pharmacological and safety tests have already been done it may not be essential to repeat all of these relative tests with the earlier and the new product should be performed to test whether both are equivalent in all critical aspects. In the case of higher standards this may be relatively simple, lowering these standards, however will be complicated to justify and requires through studies about the nature and effect of the supplementary impurities.

Analytical Development and Quality Assurance

At the beginning quality controls were carried out only with the final product with abolition of batches that did not pass the test but at the present time quality assurance covers the whole manufacturing process. A quality standard cover all that goes into the product or come into contact with the product or has the prospective to influence product quality without being noticed by the standard checks.

For starting material and the excipient used in the final formulation, analytical procedures have to be developed for the active component. These analytical tests should be competent to identify and substantiate the identity, purity, potency, stability and consistency of these materials. The process organize methods must be devised and developed to scrutinize all applicable steps of the manufacturing process in order to have sufficient control over the inborn variants and to distinguish potentially detrimental contaminants at a phase where they may be easier to mark out. The range of tests to be employed comprises methods from microbiology, immunology, protein and peptide biochemistry, genetic engineering and physio-chemical methods in addition to animal experimentation. If these exist *e.g.* in pharmacopoeias standard method metaphors have to be followed. Significant steps of the manufacturing process and analytical methods have to be validated cautiously. It will be required *e.g.* for the inactivation or abolition of potentially pathogenic micro-organism from the product.Analytically investigating specificity, sensitivity, detection limits, qualification limits, accuracy, precision, applicability and practicability under laboratory conditions and the sturdiness validates methods

Good Manufacturing Practice (GMP) principles are an additional step towards the conception of total eminence. Implementing and maintaining a GMP status requires commitment of the entire organization and special and constant attention by those who are in change of quality assurance because of the involvedness and extensive range of the subject

The organization, management structure, personal qualification and training, Standard Operating Procedures (SOPs) with suitable credentials and implementation systems ensuring their efficiency function, production, packaging, labeling, handling, testing and approval of starting materials, intermediate and final product, construction, transportation and preservation of building and equipment all these include in GMP

On the operational or technical level, GMP as well as GLP (Good Lab Practice) and GCP (Good Clinical Practice) are mainly acting. Other quality assurance standard has been established by the international organization of standardization (ISO) to presuppose eminence at the organizational level. Insufficient organizations with ambiguous tasks and responsibilities can strictly manipulate the quality of products and services.

Quality assurance fundamentals as suggested by the quality strategy ISO 9000-9004 will positively also be functional to pharmaceutical developers, manufacturers and also to sales organizations in near future.

A prescribed certification of acquiescence with these standards requires the implementation of quality assurance schemes conversing.

The subsequent aspects of a business are,

1. Administration accountability and dedication.
2. Quality assurance system covering all functions in procurement, production control of production, handling, storage, product supervision and design.
3. Identification of non-conformity and connective actions.
4. Use of arithmetic methods to develop and sustain eminence at all stages and of all actions in the product sequence.

It is predictable that acquiescence with ISO quality management standards will be mandatory for staying in business. The personnel cost of implementing and maintaining these quality assurance systems should not be under estimated.

Development and Managing Product Development

The growth of a modern pharmaceutical product is a multifaceted task. It is complicated for biological or biotechnological products because of the potential unpredictability of the biological production system and the consequential consequences. For the success of development apart from considerations of ample skills and resources, carefully and very detailed planning and controlling is vital.

The universal aims of the project and the most applicable objectives have already been decided for a reasonable and useful product outline. The objectives that provide as course points in the project will most likely address the exact indication, efficacy, and the type of product, its formulation and application scheme and also expected price limit for the production of the active component or the formulated product.

Process Manager

The accomplishment of any larger and complex process depends to a great extent upon an effective management; development process for medicinal products need process managers with a array of skills and personal qualities to manage with the very demanding job A good working knowledge in different scientific fields such as medicine, microbiology, immunology, protein biochemistry, pharmacology, toxicology, biotechnology, process development, analytical methods, formulation, and manufacturing are essential for the planning and management of a process. Process managers must be prepared to endlessly learn about these aspects. Major tasks of process manager are as follows;

Expansion of a Project Perception

Define objectives, milestones, and result points and process tasks

Scheduling of project tasks

Preparation of resources

Organization

Detection of shortages

Conciliation of use resources versus time performance

Training and initiation of major decisions

Organization and chairing of meetings

Development and stipulation of misplaced skills

Cost monitoring

Acquiescence with time frames and quality criteria

Efficient use of resources

Competence of documentation

Efficient communication among process group members

Commencement of scientific and technical report and documentation

Development reports

Presentation and justification to the management

To coordinate and control the inevitable setbacks management of skills, certain personal characteristics and psychological knowledge is required. Process managers must always think in advance, sense prospective problems before they actually appear and formulate calculate to avoid them. Process managers are always in the most outstanding position when problems are tackled. They are the scapegoat and need a high level of resistance against disappointment. Everyone will claim success.

Process managers need clear and ample organizational structures with delegation of defined duties and authority. Assets and authority should be at the process manager's outlook to react according to the needs of circumstances. For the benefit of the process, this must include the full authority to use the planned and agreed resources plus a certain degree of flexibility in special cases without the need for extensive discussions, negotiation, and approval procedures. Group member's assistance and support is essential for a project manager. Process managers can accomplish their task only by delegating work to other members of the group. If enduring pressure applied to ensure that objectives are met completely and time, this could either indicate that necessary skills are missing or that the plan was too ambitious and unachievevable. Since the individuals should know best what they can or cannot achieve in certain time, this can be avoided by a delegation of the planning details (and provision of assistance where necessary) to the functional groups and to the individual project members. If the problems remain the same, the project manager faces the very difficult task for changing the general attitude and motivation of the project group.

Managerial Structures and Impulse

An institute that focuses exclusively on only one project should not have major in-house conflicts about possessions. Above standard situation different departments or groups with different skills add to one or several projects besides their own routine work, *e.g.* in research. Somebody within these departments is nominated as project leader and continues to report to the head of his department. Moderately frequently the diverse leaders of these functional departments actually lead the project, each of them with a particular personal opinion and partiality. A common solution to accommodate the needs of development projects is the creation of matrix management structure in this project manager is appointed, who reports directly to the higher management and, as far as it concerns the project, is generally on an equal level with the practical managers who control the possessions. The power of the project manager reaches across the functional departments.

In commercially oriented organization research projects with a higher priority when development projects rarely exist, it also strengthens development activities. *i.e.* if the functional departments conduct research and development, conflicts between these two activities are usually resolved by giving the higher priority to the development project.

The steering committee acts as project obligatory, approves project plans, decides in case of conflicts and has the generally accountability for product development. The project managers should be members of the steering committee.

Across the functional departments matrix organizations offer an efficient use of resources and a better synchronization (BARTLETT and GHOSHAL, 1990). On the other hand, much time must be spent on proper planning, coordination, negotiation and finding agreements about details. The entity in a matrix organization can be caught in difficult situation and bears the disagreeable penalty if the structure does not function accurately. The inherent unsteadiness of the system may result in a situation where the individuals forget about projects and set their own priorities.

Main impulse and Displeasure Factors on the job

(a) Impulse results from,

Accomplishment
Acknowledgment
Occupation itself
Accountability
Development

Displeasure results from,

(b) Company policy and management
Administration
Association with supervisor
Work environment
Remuneration

Managerial structure and management have a noteworthy prospective for frustration among the staff and thus may severely influence the overall performance of the organization. Impulse can quite simply and effectively be created by the allocation of decipherable sections of work or independent part of a project along with ample authority to make own decisions concerning these parts. This allows the individual to assume responsibility, creates more interest in a given job and makers sure that the individual's achievements become visible and are recognized.

Impulse appears to bear a resemblance to biotechnological products, they both have a incredible prospective and are economical and highly efficient. In practice, on the other hand, their accessibility is rather restricted because their submission is more difficult than anticipated.

Chapter 9

INTELLECTUAL PROPERTY AND INDUSTRY STRUCTURE

It is observed by the historians of the biotechnology industry that despite industry legend surrounding the explosive growth of the 1970s and 1980s, neither the commercial exploitation of academic life sciences research nor the technology itself was without antecedents. The combination of underlying factors was new acting during that period to bring product development and molecular biology research closer together than ever before: ready availability of venture capital and other private funding sources; declining public support for scientific research; and changing expectations about the roles of academic institutions, as well as the technical achievements of many others. Previous links between university scientists and industry had largely preserved the restrictions between the commercial and academic spheres. These restrictions became increasingly unclear from the late 1970s.

Subsequent World War II, relative independence was enjoyed by academic science from external influences as the result of scientists themselves distributing large government grants through funding agencies such as the United States' National Research Council. Given this background, closer relationships between universities and industry is not surprising in the emergence during the 1970s and early 1980s commercialization of research particularly those in which individual scientists or their institutions expected to profit directly from the generated lively discussion among academic scientists about the possible impact of commercialization on the veracity of the research process. Concerns was expressed by scientists and observers at the beginning of the 1980s related to the prospect of dictating the direction of research with corporate interests, due to the undermining of traditional peer review mechanisms deterioration in the quality of research, exploitation of postdoctoral researchers and graduate students, financial conflicts of interest and divided loyalties, and the danger of loosing their credibility by academic scientists as impartial experts on matters of science policy.

Though surrounding circumstances have changed but many of these fundamental concerns are still applicable. For example, in the early days of recombinant DNA technology, when questions of safety were still being openly debated within the scientific community, the appearance of impartiality was necessary if scientists as a group were to succeed in resisting the obligation of legislative controls on research. Now that potential risks associated with the technology are again the subject of active debate, this time on a much broader scale, the need for impartial scientific advisers to inform the wider community is being keenly felt. Similarly, it is arguable that our current lack of detailed empirical information about risks bears out early concerns that commercialization would influence the research agenda towards short-term product development and away from more complex and longer-term problems such as determining the ecological impact of genetically engineered crops.

Questions of risk of course are, and were, highly controversial. But in their most general form, early concerns about the impact of commercialization did not depend on any particular point of view about risks or about the relative merits of different lines of research. The scientists and others expressed mostly general concern in the early 1980s was that interfering with the flow of scientific communication, commercialization will disrupt relationships among scientists and would hinder the overall progress of science.

Patenting of biotechnology inventions has increased literally exponentially, since the early 1980s, with many if not most essential tools of biotechnology research becoming subject to private possession.

If all the different owners of biotechnology patents are able to negotiate with one another so that they can put together a useful-sized bundle of rights to serve as a basis for further development, this is not necessarily a problem. But negotiations can be costly and difficult and there is the possibility that research and development will pulverize to a halt if transaction costs get too high.

It is complicated, to tell in practice, to what extent, if at all, real obstacles in biotechnology research and development on the propagation of intellectual property rights has generated.

Intellectual Property

Human beings are distinguished from animals by the intellectual property. So by means of these human beings are elevated themselves. It is common that any property moveable or immovable is to be legally protected in order to prevent it from stealing. The law of the country, named as Intellectual Property Rights, governs this property. Patents are for new inventions. These inventions are solutions for scientific and technical problems. Trademarks are useful for the consumer for the identification of the manufacturer of the product and copyrights are the rights in literary, artistic, software etc. the commercial value of any industrial product depends upon various factors such as utility, cost of production, reputation established in the trade and attractiveness of production.

Therefore businessmen and industrialists always make efforts to utilize these factors to gain competitive advantages over their rivals.

Utility and cost of production may be gained by the adoption of new inventions, attractiveness of appearance of new or original design for the shape, configuration and ornamentation of the product.

By adopting and popularizing suitable trademark for marketing the product one can gain reputation of the product in the trade, thus the use of inventions, designs and trademarks leads to enactment of legislations for the protection and regulation of the rights in such properties. So the patents for inventions, designs for industrial designs and trademark, which are concerned with industrial products, are referred as Industrial Property

Intellectual Property Rights (IPR)

Intellectual Property Rights are the legal rights in the intellectual property.

Protection of the legal rights of the intellectual property rights by the way of patents is an important tool to promote inventions and to ensure adequate returns to the investments. It is also important for technological, industrial and economic development of country.

There are various forms of IPR, known as Trade Related Aspects of Intellectual Property Rights (TRIPS) recognized by World Trade Organization (WTO)

a. Patents
b. Designs
c. Trade marks
d. Coy rights
e. Geographical indications
f. Integrated circuits
g. Trade secretes

Utility model plant variety protection patents: patents are legal rights for new inventions in scientific and technical knowledge for limited period.

The grant of patent rights has to strike balance between different conflicting interests such as, between keeping valuable technical information as trade secrets and disclosing such information between public and private interests of the inventor/ industry between excusive rights and promoting competition and particularly in developing countries, between the interests of indigenous industry and transnational corporations. Industrial Designs a design can be either two or three-dimensional. It is an idea to features of shape, configuration, pattern or ornament applied to an article. The main criteria for securing protection for a design are novelty and originality of the design.

Trade Marks

It is visual symbol in the form of a word, device or a label applied to an article of manufacture or commerce with a view to indicate the purchasing members of the public about the origin of manufacture of the goods affixed with that mark. It also distinguishes such goods from the goods manufactured by others in the trade. *i.e.* trademark enables a customer to distinguish the products of one manufacturer from that of others.

Trade mark is effective tool to attract customer by the widespread and extensive use of trade mark in public it become popular and eventually acquire an exclusive right on the mark which can be legally enforced by the owner of the mark.

Copyrights

It is right to copy and make use of literary, dramatic, artistic works, musical, cinematographic films, records and broadcast. Copyright is a proprietary right and comes into existence as soon as the work is created. It governed by the statutory laws of each country.

Plant variety protection: in developed countries it is possible to obtain legal protection for inventions using and/ or involving biological materials. Geographical indications: This indication identifies a good as originating in the territory of a country or other characteristic of the goods is attributable to its geographical origin. The TRIPS agreement requires the member countries to enact legislation for the protection of geographical indications.

Utility Models

It involves inventions in the mechanical field. The utility model system facilitates creating a culture of creativeness and builds up the mind for enhancing creativity. The degree of inventive is very small so as compared to patents and utility model law is very simple; time taken for securing the rights is short and the cost very low. Protection for living organisms: developments of technologies based on tools of biological sciences such as microorganisms to produce industrial products such as alcohol, enzymes, and antibiotics resulted in the reorganization of such tools and slowly and steadily patents covered the inventions in the areas of biotechnology.

IPR for commercial success of a product/ process the inventor has the right to commercialize the invention if others do not already patent it. The know how of the invention is very difficult to keep secret after commercialization. Therefore invention should be patented. Such protection leads to monopolies and to eliminate or reduce competition therefore IPR leads to conflict of public and private interests such as controlling the prices of essential goods, development of technology, and protection of domestic industries against foreign monopolies.

Biomedical Biotechnology

Despite the fact that, even if many industry participants still give consideration to deal with other owners of intellectual property rights are worth doing in spite of high transaction costs, from a broader perspective there are social costs associated with many of the working solutions that have been adopted within the biotechnology society.

In addition, various solutions such as building up a self-protective patent portfolio thus to improve one's bargaining position are simply unavailable to the poorest and smallest participants. In the light of this study, it is arguable that at the very least, biotechnology tool in open source licensing might symbolize a less costly or more available instrument for negotiating the patent thicket than some others that are available.

Inefficiencies associated with restricted access to intellectual property in biomedicine

Here is a synopsis, in note form only as at January 2003 and complete with voice software "typos", of Walsh, John P., Ashish Arora, and Wesley M. Cohen. 2002. The Patenting and Licensing of Research Tools and Biomedical Innovation: United States National Academies' Science, Technology and Economic Policy Board.

It considered possible impacts of the patenting of research tools on biomedical research. First, we think on whether the existences of multiple research tool patents associated with a new product or process pose particular challenges for either commercialization on or research of biomedical innovation. Second, we examine whether restricted admission to some of stream discovery perhaps sheltered by only one patents has significantly impeded subsequent innovation in the field. The access of an issued latter, at least two foundational upstream discoveries, has not yet impeded biomedical innovation considerably, but our interviews and prior cases suggest that the prospect exists and ongoing scrutiny is warranted.

Patenting of Research Tools

The patenting of research tools has made the patent scene more complex. Interviews confirmed that there are on average more patents and more patent holders and before involved in a given commercialise the ball innovation in biomedicine, and many of these patents are on research tools. In spite of this increased complication, one cannot find that socially meaningful projects are either stopped or not undertaken at all. Firms have been able to develop working solutions that allow their research to proceed. These working solutions combined taking licenses, inventing around patents, going offshore, violation (sometimes under an informal and typically self-proclaimed research exemption), the development and use of public databases and research tools, and court challenges. Besides, the members of a research community both academic and commercial researchers are a bit averse to assert their IP against one another if that means they will sacrifice the goodwill and information sharing that comes with it. Changes in the institutional environment, particularly new PTO rule.Finally, the very high technological prospect in this industry means that firms can shift their research to various less current by intellectual property claims, and hence the walling off of particular areas of research may not, under some circumstances, precise a high toll on social welfare. Those stopped and stillborn projects are not apparent, many of the working solutions to the IP complexity can impose social costs. Firms' circumvention of patents, the use of alternate research tools, inventing around all going offshore while all privately rational strategies constitute a social waste. Court challenges and even the contract negotiations themselves can also impose significant social costs. Litigation can be expensive and Norma out-of-pocket costs, represented by the effort is devoted to the matter by researchers and management, can be substantial. Even when there is no court challenge, the negotiations can be long and complex and may impose costly delays. Disagreements can and have led to lawsuit, which is particularly costly for universities and small firms. It is hard to know, however, how much contracting costs in biomedicine echo an enduring feature of IP in biomedicine, and how much is transitional. As new firms and institutions develop into owners of intellectual property, there is a costly period of adjustment as these new actors learning to manage their IP effectively.

Combined Research: Access to Initial Tools

In view about the issue of right to use, it is helpful to make a distinction research tools along two dimensions. First, it is clearly of interest how necessary a research tool is for following innovation both in the sense of whether the tool is key to successive research, and in the sense of the span of innovation that might depend on its use. Second, to what degree is a research tool rival-in-use, *i.e.* primarily used to develop innovations that will compete with one another in the marketplace. The defining feature of research tools that are not rivaling use is that the use of the research tool by one firm will not typically reduce others' profits from using it. Such tools include PCR and combinatorial libraries. From a social welfare viewpoint, a research tool that is not rivaling use is like a public could in that it has a high fixed cost of development and zero or very low marginal cost in serving an extra user. Thus, maximization of social welfare requires that the tool be made available to as larger set of users as possible. Empirical observation is that holders of IP on not rival research tools often charge prices that commit broad access, at least among firms. In some of these cases, IP holders have charged high prices to commercial clients and lower prices to university and other researchers who plan to use the tool largely for non-commercial purposes. From a social welfare viewpoint, such prices relation expands the use of the tool and is indeed welfare enhancing. On the other hand, there are cases where the IP holder cannot or does not developed pricing strategy that allows low value and academic projects access to the tool. Perhaps the greatest concern with regard to IP access tends to emerge when a research tool is both rivaling use and is potentially key to progress in one or more broad therapeutic area is. When a foundational research or is rivaling use, the IP holders often either tends to develop the technology themselves or grant exclusive licenses. As suggested earlier, special exploitation of a foundational discovery is unlikely to realize the full potential for building on my discovery because no one firm will be either capable of or able to visualize the different ways the discovery might be subjugated. The social welfare analysis of this situation is not clear-cut, however. Even though knowledge, once developed, can be shared a little additional cost and may be best exploited through broad access, it does not follow the social welfare is maximized by mandating low-cost access is such access dampens the incentive to develop for a social to begin with. Many of the same kind of working solutions that diminish the prospect of anticommons also apply to the issue of access for research. One can rationalize the failure of the IP holder to aggressively monetary infringement as a form of price discrimination that can improve social welfare.

Two vital questions are to ask when considering the effects of a given research tool patents on the growth of biomedical research. The first has to do with the specifics of the biology in question does existing scientific information provide us with many or only a few opportunities for modifying the biological system in question? Where there are many opportunities, the likelihood of a research tool patents impeding research is smaller. The second question is related with the particulars of the legal rights in question. Through an arrangement of like an suitable institutional response, one can appear to have avoided situations in which is a single firm organization users is patents to impede research in one or more broad restorative area is. Though, one must remain alert of the danger that a broad area of research could be blocked off by patents holder trying to reserve the area exclusively for itself. The question is either something systematic need to be done. One possibility is a revision of the law providing for research exemption is to better reflect the current norms and practices of the biomedical research community. It is not easy to tell when researchers non-commercial or commercial irrespective of the type of institution doing a research. So it isn't evident that society would benefit from a policy response as opposed to continuous reliance on current informal practices of violation under the informal rubric of "research exemption". It is a social waste to keep aside a large amount of money as insurance against patent infringement suits, as the firms are obliged.

While there can be some reason for optimism, future problems cannot be ruled out resulting from patents currently under assessment, new shifts in technology, court decisions or even assertion of patents on initial discovery is. One can expect that from time to time new entrants and new technologies, or those shifting from a repeated play to an endgame strategy, will create problems. Yet the system empirically observed has appeared to develop a strong combination of working solutions for dealing with these problems. The issues that these solutions can linger on and expense to workout and the results may not be best from either a social welfare or private outlook though research generally moves further. The theory of the patent system is that the benefits to biomedical advance through provision of incentives to invest in R&D. Therefore, the industry seems to be succeeding can be the conclusion of this paper, even if with some difficulty, in developing and accommodation that incorporates though the need to provide strong incentives to conduct research and the requirement to maintain free space for discovery. In the cumulative innovation model have empirically been observed to exist in biomedicine, but that in practice research is generally not obstructed. Keeping in mind these results related biomedicine rather than agricultural biotechnology, we may also look into that research goes in advance at potentially noteworthy social cost and the question is either this social cost is justified by the incentive effect of allowing patents on research tools not easy to answer. In addition, In low income public research organizations in the developing world some of these workarounds may be inaccessible, and also some of them rely on information databases and the development of public research tools, Its main aims is to encourage. One of a range of workarounds can be possible to see open source, and one that may be more accessible to poorer organizations. It also doesn't worth to say that people are able to avoid its consequences by infringing legal rights because the law is fine. The cost of revision of laws that only work when they are ignored may be high and not worth paying, but that is a separate question which is to be answered by empirical studies.

When considering the effects of a given research tool patents on the progress of biomedical research two central questions can be asked. The first has to do with the specifics of the biology in question does existing scientific knowledge provide us with a lot of or only a not many opportunities for modifying the biological system in question. There are many opportunities, the likelihood of a research tool patents impeding research is smaller. [Charles Lawson; but the human genome, for example, is effectively a standard.] The second question has to do with the specifics of the legal rights in question. Through a combination of like proper institutional response, one can appear to have avoided situations in which is a single firm organization user is patents to impede research in one or more broad therapeutic area is. However, we must stay put alert of the danger that a broad area of research could be blocked off by patents holder trying to reserve the area exclusively for itself. The question is whether something systematic needs to be done. One possibility is a revision of the law providing for research exemption is to better reflect the current norms and practices of the biomedical research community. However, is not easy to tell when researchers commercial or non-commercial irrespective of the type of institution doing a research. So it isn't apparent that society would benefit from a policy response as opposed to continuous reliance on current ad hoc practices of infringement under the informal rubric of research exemption.

Agricultural Biotechnology

In agricultural biotechnology, over the past five or 10 years a dramatic series of mergers and acquisitions have taken place that have resulted in a large proportion of elementary research tools ending up in the hands of a small number of big multinationals.

Merging with one another is one way that industry players can prevail over the costs of negotiation and build up useful-sized bundles of rights: in other words, the merger is one type of Merges' institution-forming success. But it would be surprising if such a high level of industry concentration had no adverse effects on further innovation. Moreover, although there are numerous possible explanations for the recent consolidation of agricultural biotechnology firms, some empirical evidence suggests that the changes have been driven largely by intellectual property management considerations.

Inefficiencies Connected with Controlled Admission to Intellectual Property in Agricultural Biotechnology

Among private corporations in agricultural biotechnology, an extreme form of contract cost-lowering institution has emerged. Unable to afford the costs of obtaining and enforcing patents on their own, companies have resorted to mergers, explained by Barton as the ultimate cross-license. As an outcome, a small number of powerful firms held key intellectual property rights in agriculture, largely protected from the costs of litigation by the deterrent effect of mutually assured destruction. While this conclusion may constitute institution-forming success from the point of view of the major players, the effect of this market awareness has been to raise the barriers to market entry of others wishing to participate in research and development in agricultural biotechnology, because -- apart from anti-trust laws there is nothing to discourage industry leaders from using their intellectual property portfolios against outsiders. In other terms, problems with maintaining a healthy competitive market, acknowledged by merges as a possible downside to the formation of transaction-cost lowering institutions, are apparent in the context of agricultural biotechnology.

While the major private sector players have minimized the costs of bargaining over proprietary research tools through consolidation, there is facts that other members of the agricultural biotechnology community are experiencing the effects of bargaining breakdown. As in the case of biomedical research, the problem affects different types of institutions in a different way. With respect to obtaining revenue by licensing out to other researchers, some institutions own very little intellectual property, while others (including many universities) may own substantial intellectual property portfolios, but face difficulties in relation to effective management and marketing. With respect to gaining access to research tools the clearest peculiarity is between research institutions in developed and developing countries. Researchers in developed countries are often under the false impression that they do not need to get permission to use other people's technology on the basis that they and their institutions are protected from any intrusion action by a research exemption. In many cases in the actual risk of being sued for infringement may be quite low, and the punishment imposed even after a successful infringement action relatively light, especially where the research in question is clearly non-commercial. However, as public and non-profit research institutions form closer relationships with industry and alter their activities to minimize the impact of declining external funding, the difference between commercial and non-commercial research becomes increasingly vague; thus, the risks associated with infringing uses of research tools in developed countries are likely to increase. By distinction, researchers in under developed countries are inclined to overestimate the risks associated with using other people Technology.

Nottenburg et al report a common misconception that a patent awarded in one country confers property rights in all countries. In fact, the cost of getting protection in many different countries is such that most inventions are patented in just one or a few developed countries with large markets: in 1998, the number of patents granted in the US, Europe and Japan accounted for about 80 percent of the world's patents, and it is likely that most of the remainder were also granted in developed countries. Although problems might theoretically arise with regard to technologies destined for crops grown in developing countries if those crops were subsequently exported to countries where the relevant intellectual property is protected, it has been shown empirically that exports from developing to developed countries are generally dwarfed by production and consumption in the developing world and that the value of these exports is concentrated in a few crops and a few exporting countries. Therefore, researchers in under developed countries are perhaps not seriously constrained by intellectual property considerations in the stringent legal sense. Depending on the manner in which developing countries choose to apply their obligations under TRIPs, this might soon change: in many developing countries, new patent laws are starting to come into effect such that technologies previously legally available to researchers in those jurisdictions will be protected. To a certain extent, the chief current intellectual property-related problem for research institutions in developing countries has lack of knowledge about intellectual property policy and an associated lack of capability to manage intellectual property successfully; these deficiencies stop such institutions from taking advantage of their relative freedom from legal constraints

It must be acknowledged in this connection that boundaries on access to proprietary research tools are not the only barrier to further research and development in agricultural biotechnology just as successful research and development will not itself solve the problem of food uncertainty in developing countries. Nottenburg et al point out that the real problems facing research institutions in many developing countries are a lack of local investment in science and limited experience and expertise in using and variable, as well as accessing, modern biotechnologies. C. S. Prakash also regards problems of access to proprietary technologies as just one of several factors standing in the way of applying biotechnologies in the agriculture of developing countries; adding up to those identified by Nottenburg et al, he refers to the spread of disinformation by major agricultural technology companies about matters relating to the most needy sectors of developing country agriculture, conduct which has led to a severe loss of trust. In developed countries, the real stumbling block for public and non-profit research institutions is the commercialization of research results; in many cases obtaining freedom to operate regarding intellectual property may be less of a barrier to commercialization than other factors, such as a lack of acceptance of hereditarily adapted products and uncertain government regulation. Nonetheless, members of the agricultural biotechnology community, including international aid agencies, acknowledge problems of ensuring unending access to proprietary research tools as important. Even if intellectual property issues represent only one thread in a bigger twist, extricating one thread can sometimes set the whole knot unraveling: easing problems of access to research tools could leave researchers free to work on the authentic scientific problems underlying consumer backlash and uncertain regulation, while developing country researchers have suggested that clearing the intellectual property logjam in developed countries would in itself go a long way to helping developing countries gain access to new technologies.

Research Tools in Agricultural Biotechnology

Adopting the non-profit perception, number of options for obtaining access to other people's technology, among them several which would fall into Merges' category of transaction cost-lowering institutions.

Cross licensing is the first option. Observing that this is a popular solution for deals among agricultural biotechnology. The possibilities for out-licensing by public and non-profit institutions in exchange for access to privately owned technology. For these institutions cross licensing tends to be more part of the problem than a solution. One reason is that the number and value of intellectual property bargaining chips held by most public agencies operating in or for less developed countries is small compared to those available to large private sector firms. This is somewhat due to the way public agencies have been constituted -- for example, of the approximately 670,000 accessions of germplasm held by the International Agricultural Research Centres (IARCs) of the Consultative Group on International Agricultural Research (CGIAR), approximately 520,000 cannot be used in bargaining with the private sector because they are held in trust under a 1994 agreement with the United Nations Food and Agriculture Organisation (the FAO Trust Agreement) to be made freely available to public and private sector researchers alike. It is also partly due to the complete difference in resources available to the public and private sectors. The second option for non-profit and public research agencies is to obtain low-cost, research-only licenses from owners of investigate tools. While this option does have some advantages for the licensee, encounter of difficulties often outweighed them if the research project succeeds and the researcher wishes to commercialize or otherwise disseminate the results: at this stage the researcher is at a disadvantage in bargaining for a broader license, as the costs of the research have already been incurred. Market segmentation strategies is the third option to employ based on geography, field of use, length of license, or other criteria, the idea being that the owner of a research tool will be more willing to grant access to the tool if the owner's income are not likely to be threatened by its use. A linked option is to negotiate license fees, which are scaled according to the licensee's ability to pay, or the amount of profit the licensee expects to gain from the use of the research tool. For market segmentation of pharmaceuticals with large potential markets in both developed and less developed countries which would effectively require patent applicants to choose between enforcing their patents in developed countries or developing countries, but not both; they suggest that plant biotechnology patents could be included in this scheme. The difficulty for public and non-profit research institutions in relying on market segmentation strategies and scaled fees is that successfully negotiating agreements that represent these strategies requires significant expertise and other resources.

The fourth option is to form links with owners of research tools through mergers or joint ventures. Noting that for many public research institutions privatization through merging with a private company is neither viable nor desirable, they suggest that joint ventures between public research institutions and industry can bring noteworthy remuneration. Though, such relationships can be controversial, for example because of concerns about conflicts of interest and undue influence on the part of the industry partner on the conduct of research or dissemination of results. From the private firm's position, the condition of such support may be motivated by the prospect of close contact with the pioneering activities and expertise of the public sector institution; for example, a firm might sponsor a research program in a university laboratory in order to be able to recruit graduate students trained in skills that are highly valuable to the firm. Another incentive for a company to provide such support might be the chance of obtaining tax deductions in exchange for unused, perhaps ineffective, technology. Some private firms appear to have taken this approach to the extreme by making well-publicised free donations to public research institutions of technology to which the private firm does not in fact have an special right, for example because the relevant patents have expired or are otherwise invalid. Genuine cost-free licenses may sometimes be available for technologies relating to crops which are not of any commercial interest in developed countries, but in such cases the private firm would like to be protected from culpability, loss of status or accountability for misuse of the technology in countries that lack effective regulatory oversight. They do not mention liability for harm caused by proper use of the technology, perhaps because even though current biotechnology regulation even in developed countries does not in fact exclude the risk of such harm, it would be almost impossible for victims of harm to prove liability. One such option is for non-profit or public institutions to partner with independent developers of research tools who may sponsor the creation of substitutes for existing proprietary tools that is, inventing around intellectual property rights, a strategy which is beyond the resources of many non-profit institutions acting on their own. One such independent institution is the Center for Applications of Molecular Biology in Agriculture (CAMBIA), which develops agricultural biotechnology research tools unfettered by restraining proprietary claims; these tools are made available to other research institutions on an ability-to-pay basis. Another choice which relies on taking advantage of common interests within the public sector is for non-profit and research institutions to get together with powerful allies

For potential users of research tools to get a joint grant of freedom to operate in certain markets from all holders of significant intellectual property rights, through the owners' membership of a patent pool which bundles members' intellectual property rights and licenses the bundles to third parties. This option may not be practicable as a long-term solution to the access problems experienced by public and non-profit institutions because the operation of the patent pool, and thus the availability of particular bundles of licenses, would be beyond their control. A more promising option is for members of the agricultural biotechnology community generally (or a subset of that community, including public and non-profit institutions) to establish some kind of collective rights organization which could perform a range of functions, including that of assembling useful bundles of intellectual property rights belonging to members and possibly also to non-members.

A number of voices have advocated the creation of such an institution in the field of agricultural biotechnology. Peter Drahos proposes a global bio-collecting society (GBS) to help overcome problems of uncertainty and enforcement that confront life sciences companies and native groups constricting over the use of knowledge relating to plant material. The proposed GBS would be analogous to collecting societies that already exist in the copyright area, but would differ in two important respects. First, it would be a single international organization, because this would promote greater transparency and would serve the interests of indigenous groups better than having many national societies. Second, protecting the interests of the indigenous intellectual property owners would be a primary but not an absolute duty of the proposed GBS: it would be chartered to serve the broader purposes specified in the Convention on Biological Diversity and perhaps also the International Undertaking on Plant Genetic Resources. Drahos suggests that a GBS could be set up as a private organization outside the framework of any treaty cooperation, that it could be funded by the World Bank, and that membership should be open to companies as well as local groups. Membership would be optional: if the GBS proved helpful to contracting parties, it would automatically attract and keep enough members to continue functioning. Drahos analyses the advantages of a GBS for indigenous groups and members of the life sciences industry and concludes that both groups would have adequate incentive to participate.

Drahos proposes five functions for a GBS. First, it could be a repository for classified community registers of indigenous knowledge. Companies could access the register to the extent of identifying which group had made an entry and could then approach that group directly. Second, it could smooth the progress of negotiations between indigenous groups and third parties, including by keeping a register of independent legal experts willing to assist indigenous groups in such negotiations. Third, it could monitor the global use of indigenous knowledge by conducting regular checks of patent applications or by keeping information about the use of licensed knowledge. Fourth, it could help to resolve disputes. Drahos suggests a standing committee, which would openly examine the conduct of parties to dispute and make recommendations. These recommendations would not be legally binding, but any party who ignored the committee's recommendations can be barred from the GBS and would be widely shamed. Drahos emphasizes that for this approach to be effective, the members of such a committee would need to be, and be seen to be, independent and impartial. Finally, a GBS may have some kind of standard setting function. For example, it could bring representatives from industry, national governments and indigenous groups jointly to develop an authoritative code of conduct for the negotiation of a biodiversity prospecting contract.

Others, like, proposals related to the establishment of a collective rights organization for international agricultural research generally. The most highly developed of these proposals envisages membership being limited to public and non-profit institutions, but some discussions have also canvassed the participation of industry. One group of scholars working towards a detailed model of an intellectual property clearinghouse for international agricultural research is located at the UC Berkeley Center for Sustainable Resource Development.

An Intellectual Property Clearing House for International Agricultural Research

(a) Need for a Collective Solution

In a 2001 article in Nature Biotechnology, Graff and Zilberman sketch out the need for a shared solution to the tragedy of the anticommons in agricultural biotechnology and introduce the concept of an intellectual property clearinghouse. Graff and Zilberman begin by noting that the efficiency of patent system turns on two factors: the definition of what is patentable, which demarcates private property from the public sphere, and the effectiveness of mechanisms available for trading in patent rights. Therefore, any appraisal of the patent system can focus on the scope or definition of what is patentable or on issues surrounding the exchange of patent rights (or both). Graff and Zilberman argue that by focusing on the exchange of patent rights, significant improvements in freedom to operate can be achieved no matter what shape patent policies end up taking.

Graff and Zilberman spot the existence of a tragedy of the anticommons in agricultural biotechnology, asserting that the current fray of agricultural biotechnology patenting has two effects: first, the interdependent or complementary nature of intellectual property rights in this area is being lost in the section among multiple private property claims, and second, difficulties in negotiating and managing access to needed intellectual property saps the power of private incentives to innovate. The cumulative result, they say, is a dampening of research and innovation productivity a "tragedy of the anticommons".

Graff and Zilberman consider the options both for individual companies pursuing freedom to operate within a overcrowded intellectual property landscape and for government intervention to solve the problems just mentioned. Both may involve the formation of collective rights organizations, a subset of Merges' "transaction cost-lowering institutions". Graff and Zilberman points out that private collective rights organizations have historically been more economically efficient than government-invoked organizations, sometimes providing substantial savings for whole industries. Further, they argue that advances in information technology have created new tools for intellectual property information management and marketing that open up new organizational possibilities for collective intellectual property rights organizations, suggesting that problems that have plagued past attempts to lower transaction costs in this area (such as online intellectual property informatics and patent exchange services) can now be overcome.

Like that of Drahos' global bio-collecting society, the goal of the intellectual property clearinghouse proposed by Graff and Zilberman would be broader than only to serve the interests of intellectual property owners. They believe that a clearinghouse could serve as a focal point for the management of public sector intellectual property related to the agricultural life sciences, to enable the widest possible commercial adoption of related technologies and to promote access for research and humanitarian use. Its purpose would be to reduce transaction costs and other market failures that hinder the exchange of intellectual property, creating pathways through the patent thicket and giving access for researchers to, and commercial freedom to operate with, proprietary biotechnologies. Such an intellectual property clearinghouse would, it is argued, have a range of positive effects. It could serve to level the playing field and free up agricultural research. It could reduce the drive for firms to consolidate in order to obtain complete in-house control of complementary technologies and enable them to improve their product design, as modernization would be less constrained by current intellectual property holdings. It could help to move proper technologies out into regional and applied agricultural research systems around the world and provide incentives and means for current outside players to strengthen their agricultural research capacities. Finally, it could help agricultural research define and maintain a healthy balance between public and private forces and grow along a more proficient, safe and beneficial path.

(b) Essential Functions of an IP

(i) *Clearinghouse*

An intellectual property clearinghouse would perform three essential functions. The first would be to identify all related intellectual property claims over a given technology and indicate the amount to which the technology is available for licensing; this basic information function would compose a platform on which more detailed services could be built. The second function would be to facilitate negotiations, and the third would be to keep an eye on and enforce negotiated agreements. In other words, the three essential functions of an intellectual property clearinghouse would be designed to reduce the three kinds of transaction costs identified by Long: search, negotiation and enforcement costs.

The second function of an intellectual property clearinghouse, like the first, would start by improving on services that already exist but are not entirely satisfactory. Some recently founded online patent and technology-licensing services have pioneered information technology tools designed to assist negotiations by matching prospective parties to a transaction and providing pricing information. Though hopeful, these services are limited in their power to manage multi-party, multi-patent transactions and, like most existing IP informatics data sources, they have not yet achieved critical mass with respect to the specific intellectual property relevant to agricultural biotechnology research. In addition matching parties and establishing prices or pricing indicators, the staff of an intellectual property clearinghouse could negotiate with technology providers to allow multiple complementary patents to be bundled into "micro pools" and sublicensed to users under single contracts; such multi-patent licenses could be constructed to provide access to "research tool boxes" containing all the platform technology required for a particular agronomic or plant system. Staff of a clearinghouse would also actively pursue flexible licensing strategies, including customized packages for particular licensees. By acting as an honest and reliable broker, an intellectual property clearinghouse could facilitate negotiations between parties who might have problems of trust or confidence due to cultural barriers or negative past experience. It would weight fair trading, avoiding reach-through agreements and other restrictive licensing requirements, offering scaled fees determined by ability to pay and renegotiating licenses with each change of intellectual property ownership. Because bio safety considerations strongly affect the value of a technology, keeping track of which components of a technology system have been approved for which uses in which countries is an important part of conducting biotechnology negotiations; an intellectual property clearinghouse could help by maintaining data about the current regulatory approval and bio safety status of new technologies in different countries.

(ii) *Monitoring and Enforcement*

The third essential function of an intellectual property clearinghouse would be to help reduce enforcement costs by contributing mechanisms for monitoring fulfillment with negotiated agreements. It could also perform a dispute resolution function at any stage from negotiation onwards, possibly modelled after Drahos' suggestion in relation to the GBS of an independent standing committee. Depending on the nature of the dispute management system adopted, there might be a capacity for providing external review of license agreements. The intellectual property clearinghouse could also offer practical services such as making arrangements for the consignment of necessary materials between parties or electronically distributing research data.

Once the clearinghouse became effective at performing its basic functions, it could extend its operations in whatever direction its members saw fit, such as IP policy development. There would, of course, be some fundamental constraints. For example, at a recent industry, academia and international development round table to discuss the possibility of an intellectual property clearinghouse, the possibility of working out general deployment agreements with companies and universities for the application of agricultural technologies in all developing countries was raised, but dismissed because it was thought that such agreements would be too inflexible to deal properly with heterogeneities within and among developing countries, differences among technologies and the varied interests of technology providers. Similarly, participants in the round table discussion considered that an intellectual property clearinghouse would be better at providing access to research tools -- "enabling" or "process" technologies -- than to specific feature or product technologies, because process technologies raise fewer ongoing regulatory, liability or other product stewardship issues.

(iii) *Summary of Vital Functions*

In a recent study directed specially at the probability for conglomeration among public and non-profit research institutions in agricultural biotechnology, Bennett et al summaries the services that could be performed by a public sector intellectual property clearinghouse by dividing these services into two tiers. The first level, designed to decrease the cost and worries of intellectual property information, corresponds to the first essential function described above, while the second level, designed to reduce the transaction costs and uncertainties of intellectual property licensing and to create new commercial opportunities, corresponds to the second and third functions and involves operations similar to those of a patent management company or patent licensing organisation such as a patent pool. Bennett et al propose a series of steps for establishing the services to be provided by an intellectual property clearinghouse to make certain that each new service builds on the infrastructure created by already established services. The advantage of this approach is that it helps to focus efforts to set up a clearinghouse in a logical sequence, but that if any of the more elaborate proposes services proves unworkable or unnecessary, the more basic services can still be made available.

First Level:

1. Licensing status information assortment
2. Regulatory information collection
3. Collection and investigation of the available public sector intellectual asset base
4. Distribution of licensing information
5. Proliferation of licensing standards and best practices
6. Advisory consulting and professional training

Second Level:

7. Patent sublicensing through an incorporated legal entity that could take on options, licenses or assignments of IP from member and nonmember IP providers and sublicense to users.
8. Patent pooling -- rights obtained from multiple IP providers that are acknowledged to cover complex technology systems could be prepackaged into patent pools to be sublicensed in single dealings
9. Practical IP management.

This final step would involve the classification of new opportunities arising out of recognized services. Bennett *et al.* give examples: shopping lists of needed technologies to be fulfilled by the clearinghouse approaching rights holders and seeking sublicenses; updating and optimizing of pools on an ongoing basis; adding new and phasing out pools; obtaining key technologies from private sector sources to round out and assemble better sublicensing opportunities; and active license marketing of unlicensed technologies of participating IP providers so as to increase the attractiveness of those technologies. Eventually, it is suggested, the intellectual property clearinghouse could become proactive in research and development by interfacing with grant agencies to help identify areas where technology bottlenecks block freedom to operate.

(c) Potential Users of an IP Clearinghouse

Bennett *et al.*'s proposal is for a clearinghouse whose main members would be organizations involved in agricultural research and development in the public sector, including government research institutes, public and private universities and non-profit organizations, although the proposal does not rule out the participation of commercial firms if they are prepared to agree to the public mission of the clearinghouse. Bennett et al also visualize the possibility of scaled membership and of allowing nonmembers access to a division of the services provided to members.

The advantage of developing a model which does not depend on the participation of large commercial firms is that it allows those organizations whose activities are most mired by the current "tragedy of the anticommons" in agricultural biotechnology, and who share related missions, to settle their differences and improve their situation before tackling the more difficult problem of negotiating with organizations whose mission is fundamentally different. However, there is no reason in principle to keep out any particular sector of the agricultural biotechnology research and development community from a collective rights organization. As was established at the round table discussion mentioned earlier, the success of such an institution must be measured in terms of the system as a whole, not by suggestion to the interests of any particular participants.

In reality, more general discussions regarding the establishment of an intellectual property clearinghouse have revealed incentives to participate for all members of the agricultural biotechnology community. Graff and Zilberman suggest that those already actively patenting in agricultural biotechnology are the most likely initial participants both as the main suppliers and the most active users of proprietary technologies. In this category would be leading multinationals, some universities and smaller laboratories and biotechnology firms. From the industry leaders' viewpoint, licensing, even outside the oligopoly, may sometimes be more profitable than special use, provided the firms involved can protect themselves from legal responsibility, negative promotion and use of the technology in competition with the firm's paying customers. Universities would benefit from any system which allowed them to license currently unlicensed and unused technologies, or to grant non-exclusive licenses where they currently have no alternative but to license absolutely. Like universities, smaller laboratories and biotechnology firms are often in possession of individual technologies which would be more valuable as part of a package; a clearinghouse would benefit them by providing a secure outlet for their technology or a source of the complementary pieces they need to commercialize their discoveries themselves.

In addition to these players, Graff and Zilberman point to the survival of many others who are not currently patenting but would benefit from a better functioning intellectual property market. Included in this category would be farmers, agricultural cooperatives and growers' associations; medium and small-scale seed enterprises and nurseries in developed countries and national seed companies of developing countries; many universities, including many land grant and public universities in the US and elsewhere; the CGIAR's International Agricultural Research Centres; National Agricultural Research Services (NARS) of developing countries; and non-government agricultural development organizations. These players would benefit from an international intellectual property clearinghouse for agricultural biotechnology by being enabled to license in currently unavailable technologies at affordable prices and on reasonable terms, that they would be encouraged to develop and license out their own inventions on reasonable terms and that over time this would encourage the development of agricultural research capacity within those organizations. The possibility of achieving this latter result through the operation of an intellectual property clearinghouse could also attract non-profit funding agencies as participants (Catherine Ives gives the Rockefeller Institute and USAID as examples).

(d) Structure and Governance

An intellectual property clearinghouse would be a private institution whose members would be linked to one another through common contractual commitments. Members would be anticipated to take part actively, give information and possibly also to provide financial support in the form of a membership fee which could be scaled according to the type of organization. (Apart from this suggestion, the question of how the clearinghouse would be funded has been left open in current deliberations.) Membership of the clearinghouse would not be obligatory, with members self-selecting according to their interest in promoting its goals; these would need to be carefully defined, especially with respect to the relative priority to be given to for-profit and caring activities. The technological scope of the clearinghouse would be limited to life sciences applications in agricultural crops, and should be relatively narrow, perhaps even focusing initially specifically on areas where the tragedy of the anticommons is particularly severe, such as the development of subsistence crop varieties. With respect to jurisdictional scope, first-level information services should be global; second-level services might need to be organized on a country-by-country basis, though eventually expanding to cover all intellectual property jurisdictions.

Debate about the prospect of an IP clearinghouse have left other issues of structure and governance open, but have particular key design criteria. Specifically, participants have emphasized that self-reliance and faith are crucial to the success of any effort to promote cooperation between individual technology stakeholders, both in developed and developing countries. To win and retain the self-confidence of members and prospective members, the clearinghouse must have clear and decisive governance (one participant suggested this meant non-consensus governance mechanisms). To maintain trust, the organization should be independent of each of its members.

One way to achieve this independence might be for the clearinghouse to be governed by a Board of Directors drawn from member organizations, and/or external directors of adequate number to assure the impartiality and objectivity of the Board. There are many possibilities: what matters is that the final shape of the institution reflects the needs and desires of all its members or prospective members and that financial and governance structures are both correctly distributed among members and obvious, to avoid any problems arising out of conflicts of interest or the formation of destructive coalitions. The clearinghouse must not be supposed as a technology users' club or a technology sellers' marketing tool: this is important both to be a focus for and keep members and to avoid falling foul of competition (anti-trust) laws.

Richard Jefferson of CAMBIA repeated an important point about the design of any joint rights organization for agricultural biotechnology research in this framework: that terms such as "biotechnology", "developing country", "public sector" and "patent" are all generalizations whose detailed meanings will need to be broken down and clarified as the terms of a clearinghouse are spelt out.

The 2001 industry, academia and international development round table already referred to ended with the following recommendations from the floor pertaining to suitable next steps in the design of a collective rights organization.

The first recommendation was to define in person intellectual property-related problems clearly and concisely and set parameters for solving these problems. The second was to search for input from the budding clients of an intellectual property-clearing house in agricultural biotechnology in order to assess what element they would like to see in its design and how they would expect to benefit by participating. It was imagined that this input would be required through interviews, focus groups and case studies among agricultural researchers in order to identify actual needs for intellectual property capacity building, information services, patent bundling and pooling, patent exchange and other (not yet specified) mechanisms. In this respect, the round table acknowledged that special care should be taken to assess the needs of the national agricultural research systems of developing countries, which were largely unrepresentative at the discussion.

The third recommendation was to focus on access to information, asking what information would be most useful to whom, what explanation and analysis should be offered with it and what opportunities might exist for partnering openly available information resources with private information tool providers. The fourth recommendation was to develop partnership arrangements within particular sectors of the research and development community: for example, public institutions and universities could pool or unite intellectual property portfolios based on the identification of equally complementary technology components.

The fifth recommendation to explore possible strategies for meeting the intellectual property needs of different sectors with common mechanisms. In particular, the round table considered that research and development relating to specialty crops and subsistence crops for developing countries encounter similar problems, so that needs in these areas should be explored simultaneously in case the same solutions were appropriate to both. On the other hand, participants in the round table acknowledged that there is a variety of needs which may require a variety of different solutions: the sixth recommendation was that those involved in developing an intellectual property clearinghouse should expect to take separate steps to reflect different parties' interests.

The seventh recommendation was that potential clearinghouse mechanisms should be pre-tested with experimental studies or simulations. The eighth was to identify in what specific technologies, particularly what research tools, an intellectual property clearinghouse should deal, and to determine what conditions the owners of those technologies would wish to specify. The round table noted that, depending on the immediate goal of the clearing house, it will be necessary to identify specific technologies in which freedom to operate is required, to determine the combinations of patents under which they are practicable, and to approach the owners of the rights represented in those patents to negotiate terms of use.

The ninth recommendation from the floor of the round table discussion was to focus on organizational questions of how intellectual property clearinghouse organisations could be structured and funded. Questions raised included exactly what form a clearing house institution or network of institutions should take; whether the institution should be a single organization with a commercial arm and a non-profit arm; whether separate services should be recognized by concerned parties in their own sectors which could then represent them and negotiate on their behalf with others; and whether a central hub to such a network of sector-specific clearinghouses could provide generalised services such as intellectual property information listings and flows of royalty payments. In terms of funding, the round table considered whether private investors would be interested in backing such a venture, and if they were, whether the public sector would view such backing as legitimate.

The tenth and final recommendation was to get beyond the generalities and assign a smaller subset of people to start work on specifics.

Research and development in agricultural biotechnology depend heavily on access to multiple research tools. One reason is that most agricultural biotechnologies are actually packages comprising multiple mechanisms. Transformation technology the means by which foreign genes coding for desired traits are included into a plant genome, allowing the regeneration of whole genetically engineered plants from the transformed tissue is a case in point. An essential tool in both commercial crop development and experimental plant biology, transformation requires access to specific gene sequences and functional information, to a range of enabling technologies (including gene introduction methods, promoters and selectable markers), and to germplasm or cultivars into which the novel genes can be incorporated.

Another reason why scientific exchange is especially important to the progress of research and development in this field is that agricultural biotechnology is not a single authority: it combines resources from many areas of biology, including crop genetics, breeding, agronomy, pest control and agroecology. For these reasons, innovation in agricultural biotechnology is both growing, in the sense that each invention builds on previous inventions, and corresponding, in the sense that each invention contains elements derived from more than one source.

Not only does research and development in agricultural biotechnology rely on access to multiple research tools, but also these tools are increasingly subject to proprietary controls. Changes in intellectual property laws mentioned in the previous chapter have strengthened protection for inventions in agricultural as well as biomedical biotechnology; stronger protection has made molecular biological techniques more profitable and therefore more widely used, which in turn has increased the demand for safeguard. Legal means of protecting intellectual property in agricultural biotechnology include patents, plant breeders' rights (in the US, Plant Variety Protection Certificates), trademarks, geographical indications, trade secrets and contracts, the first two categories being the most important. (The use of intellectual property in agricultural biotechnology can also be controlled by technical means, such as hybridisation and genetic use restriction technologies -- GURTs -- which make seed unsuitable for replanting or suppress the expression of introduced traits in saved seed. While legal rights are established by national legislation and court decisions, in practice their content is determined by international agreements, which in recent years have further encouraged the proliferation of strong intellectual property rights by requiring national governments to meet certain minimum standards of protection. As a result of these trends, research tools in agricultural biotechnology are subject to numerous overlapping proprietary claims. Depending on the complexity of a product, its development may involve the use of dozens of proprietary research tools; an often cited example is that of GoldenRiceTM, a genetically engineered rice variety developed using approximately 70 different patented technologies.

So researchers in agricultural biotechnology must coordinate numerous unequal property rights in order to obtain an effective right of use; this means the first condition fixed by Heller and Eisenberg as necessary to the creation of a tragedy of the anticommons is fulfilled. There is also evidence that transaction costs associated with obtaining freedom to research and to commercialize the results of research in this field are increasing. In fact, streamlining access to patented technologies appears to have been a key motivation behind consolidation of a number of private agricultural biotechnology firms in the 1990s. High transaction costs associated with licensing intellectual property rights in agricultural biotechnology have been reported to result from uncertainty of patent validity, excessive breadth of patents, conflicting claims of patents, difficulty of identifying valid licensors, the costs and slow pace of litigation and concern over liability, brand image and externalities control; in some cases, owners have been simply unwilling to negotiate with potential users. In addition to the costs of license negotiation, the costs of determining whether it is even necessary to obtain permission are important. Barton suggests that as a result, individuals and organizations may be too conservative in their use of proprietary technologies, thereby placing unnecessary restrictions on research and development. Proof of high transaction costs means that Heller and Eisenberg's second condition is also fulfilled.

Thus, it is visible that an assessment of bargaining behavior in the field of agricultural biotechnology could generate useful data for comparison with existing studies of the emergence or failure to emerge of Merges "transaction cost-lowering institutions" in the face of a possible calamity of the anticommons. Agricultural biotechnology is particularly useful as a point of comparison with biomedical biotechnology, the field cited by Heller and Eisenberg as exemplifying anticommons tragedy. This is because these fields are closely related in terms of both technology and the types of institutions participating in research and development in fact, there is important overlap in relation to particular institutions such as universities which are active in both fields and yet they are distinct in that they are funded in a different way, profitable products are aimed at different end consumers, and their research communities comprise separates markets or audiences for research tools. It should therefore be possible, by studying bargaining over proprietary research tools in agricultural biotechnology, to observe some of the issues that have become familiar to biomedical researchers playing themselves out in a different background. In addition, any solutions to bargaining breakdown that may emerge from a study of agricultural biotechnology might turn out to be applicable (with appropriate modifications) in the biomedical field as well.

A Stark Illustration of the Effects of Commercialising "Public Good" Research

Above and beyond providing an evaluation with other areas of research and development, agricultural biotechnology is of interest in its own right in relation to issues arising out of the commercialization of scientific research. There are a number of aspects to this reason for studying agricultural biotechnology. The first is that in agriculture, as in medicine, the stakes are high: any obstruction to the progress of research and development is likely to have serious cost for social welfare. In this regard it is necessary to recognize that the application of biotechnology in agriculture is highly controversial. on the other hand, at least to the amount that the biotechnology debate turns on scientific questions, people on both sides have reason to care about the development of agricultural research and to be concerned about the prospect that difficulties of access to proprietary research tools could strangle that research.

On one hand, proponents of the development and application of genetically customized crop varieties argue that agricultural biotechnology has the potential to solve the problem of world hunger. Hunger leads to extreme suffering and at present affects an unacceptably large proportion of the world's population. Its causes are compound, but they include poverty and lack of access to affordable staple foods. Proponents of genetic engineering point to the success of the "Green Revolution" of the 1950s and 1960s in increasing food production, but acknowledge that the technology behind the Green Revolution especially its heavy dependence on chemical fertilizers and pesticides is environmentally untenable. They wrap up there is a need for a second Green Revolution in which existing crop varieties are replaced by high yielding, nutritionally enhanced, pest- and disease-resistant genetically engineered plant varieties. From this perspective, any impediment to the progress of research and development in agricultural biotechnology prolongs the suffering of hungry people, especially in less developed countries. It is even debatable that a full-blown tragedy of the anticommons in agricultural biotechnology would be more disastrous than in biomedical research, on the basis that the mitigation of hunger could by itself be expected to lead to greater improvements in public health than any medical advance.

Opponents of genetically customized crops, on the other hand, are doubtful about the role of biotechnology in increasing food security. Instead of seeing genetic engineering as an environmentally sound option to existing Green Revolution crops, they view the release of genetically modified plants as a potential environmental and public health disaster. In spite of some attempts by proponents of biotechnology to cast those who object to its use in agriculture as irrational opponents stand on sound scientific ground when they differ that substantial further research is needed in order to identify and assess possible risks connected with the release of genetically modified organisms into the environment. So from this perspective too, free access to research tools by scientists studying the effects of genetic engineering in agricultural systems is essential.

Whichever position one takes with respect to the promises and risks of agricultural biotechnology, it matters who has access to research tools: private firms are less likely than public or non-profit organizations to be interested either in developing the kinds of crops which might be suitable for addressing food insecurity in the developing world, or in research which might depict serious risks coupled with genetic engineering. Traditionally, agricultural research, including research leading to the development of Green Revolution crops varieties, has largely been conducted in the public and non-profit sector, but organizations operating in this sector have generally failed to remain with the changes imposed by developments in intellectual property law and other aspects of commercialization, and recent declines in research funding in agricultural science have impacted disproportionately on public sector institutions. In terms of social welfare, it is important to stay away from a catastrophe of the anticommons in agricultural biotechnology because if bargaining over proprietary research tools in agricultural biotechnology fails, it is the non-profit sector which will be left without the ability to maintain research and development. (In fact, concern that this is already happening underlies much of the current opposition to applications of genetic engineering in agriculture. As Blakeney has observed, objections relate not just to the technology itself (either because of environmental or health risks or on moral or spiritual grounds) but also to risks arising from the particular social settings (which include intellectual property laws) in which the technology might be deployed. The prospect that non-profit research and consequently its intended beneficiaries, the poor and hungry, could be shut out of the biotechnology revolution, is an example of the latter class of objections. Thus, a disaster of the anticommons happening in agricultural biotechnology would be a particularly stark illustration of the adverse effects of commercialising "public good" scientific research.

In this context it is worth noting that there are two aspects to a prospective tragedy of the in agricultural biotechnology. The first relates to the problem of accessing information and biological materials necessary to the conduct of research. The second relates to the problem of ensuring freedom to commercialize the results of research conducted using proprietary research tools. The term "freedom to operate" is occasionally used loosely to cover both research and commercialization, but more often it refers to commercial freedom to operate, and that is the sense in which it will be used here. (Freedom to conduct research is sometimes referred to as "freedom to innovate", but this is puzzling because it conflicts with the common use of the word "innovation" to refer to bringing a new product to market.) In spite of the conceptual feature, both types of freedom are important to public and non-profit research institutions in agricultural biotechnology because in an environment where funding is hard to obtain, many institutions rely on the capability to commercialize some of their research results in order to support their other activities. Adding up, owning intellectual property rights even in inventions which are not planned to be sold for profit gives these institutions a positive bargaining chip in their dialogue for access to research tools owned by others.

The phrase research and development (also R and D or R&D) has special commercial implications apart from its conventional coupling of research and technological development.

In general, specialized units or centers belonging to companies, universities and state agencies conduct R&D activities. In the background of commerce, "research and development" normally refers to future-oriented, longer-term activities in science or technology, using similar techniques to scientific research without pre-determined results and with broad forecasts of commercial yield. Statistics on organizations devoted to "R&D" may express the state of an industry, the degree of competition or the attraction of scientific movement. Some common measures include: budgets, numbers of patents or on rates of peer-reviewed publications. Bank ratios are one of the best measures, because they are constantly maintained, public and reflect risk. In the U.S., a typical ratio of research and development for an industrial company is about 3.5% of revenues. A high technology company such as a computer manufacturer might spend 7%. Although Allergan (a biotech company) tops the spending table 43.4% investment, anything over 15% is remarkable and usually gains a reputation for being a high technology company. Companies in this category comprise Big Pharma, such as Merck & Co. (14.1%) or Novartis (15.1%), and engineering companies like Ericsson (24.9%).1 such companies are often seen as poor credit risks because their spending ratios are so strange. Normally such firms thrive only in markets whose customers have acute needs, such as medicine, scientific instruments, safety-critical mechanisms (aircraft) or high technology military armaments. The extreme needs justify the high risk of failure and accordingly high gross margins from 60% to 90% of revenues. That is, gross profits will be as much as 90% of the sales cost, with manufacturing costing only 10% of the product price, because so many persoanl projects succumb no vulnerable product. Most industrial companies get only 40% revenues. Commonly the largest technology companies not only have the largest technical staffs, but also manage them most successfully. On a technical level, high tech organizations explore ways to re purpose and repackage advanced technologies as a way of amortizing the high operating cost. They often reprocess advanced manufacturing processes, costly safety certifications, specialized fixed software, computer-aided design software, electronic designs and mechanical subsystems.

Chapter 10

SUCCESS STORIES IN BIOTECH

The existence of successful industrial examples represents a strong incentive to found new biotech industries. The presence of such industries, indeed, scientists, showing them how to create and run a company in the sector. moreover, it may present a key driver in localization choices of large industries

Genentech, founded in 1976 by the venture capitalist Robert A. Swanson and the biochemist Herbert W. Boyer, resulted the first established firm of modern biotechnology. Genentech started its activities near San Francisco, and represented a milestone in creating the today's U.S. largest biotech cluster (the Bay area Cluster) and in inventing the whole new biotech industry, transferring a brilliant scientific discovery in an effective business model.

Immunotech, created in 1982 by some researchers of the Marseilles Luminy Immunology Centre, represented a major event in the French academic and research environment because this type of spin-offs was (at that time) totally uncommon. The venture turned out to be a success and generated a lot of interest and attention both from the business and the academic side.

The Heidelberg Technology Park, the successful example of Lion Bioscience strongly contributed to the development of the area.

In some cases, even the public support to high potential DBFs in their initial stage has the deliberate objective to create a successful example, thus speeding up the development of the whole sector in the area.

Celltech, for example, based in Slough (UK), was founded in 1980 with the National Enterprise board (a governmental institution for the industrial policy) as its major shareholder, in order to create the first biotechnology company in the UK capable of exploiting discoveries in the biological sciences at British universities. Celltech had exclusive rights to inventions emerging from MRC's (Medical Research Council) Laboratory of molecular Biology in Cambridge.

Attraction of New Sites of other Industries

In order to attract quality of new sites from other industries, the following practices can be highlighted:

- The provision of economic incentives for the establishment of new sites, creating an annual budget managed by the local government and assigned with regard to the "quality" and the number of new sites;
- The provision of tax credits for research activities in the first years of establishment. Different from direct incentives, tax credits represent a cost reduction for new sites and are available only for research centres. In some cases, tax incentives may be extended for example to venture capitalists, stimulating the investments in biotech within the cluster.

Science and Technology Plan (Taiwan)

Since 1984 Taiwan is developing a science and Technology Plan focused on biotechnology. On of the main objective of the Plan is to establish Taiwan as a link in the international community of R&D and the commercialization of products, presenting Taiwan as a service and manufacturing centre for biotech. Five new science-based industrial parks were created, offering favorable rents and shared machineries (conditions extended also to foreign companies). To this aim the government created a fund of €3.1 billion. Another fund (Development Fund) of €57.5 million supports large scale plant development. Up to 2002, 19% of the Development Fund has been used to support large projects of foreign companies and the creation of around 14 new biotech companies.

Moreover, biotech investments have been stimulated by tax incentives: venture capitalists investing in biotech business are allowed a 25% tax rebate on capital gain.

Arknsas' General Assembly passed incentives during the 1997 legislative session that are tailored specifically to biotechnology firms. A summary of the incentives includes:

1. A 5% tax credit on the cost of biotechnology facilities;
2. A 30% tax credit on the cost of employee training when conducted through a higher education partnership;
3. A 20% credit for the cost of qualified biotech research that exceeds the cost in base year 1996 (*i.e.* the 1996 tax tear). Any unused credit may be carried forward for 14 years after the tax year in which the credit originated. Cumulatively, these credits can create a significant tax savings for companies that invest in the state and reduce the cost of doing business.

Integration Among Industrial Actors

The practices analyzed regarding the diffusion of the networking culture in the scientific base can be easily adapted to the case of the industrial base; moreover, the following practices can be highlighted:

- Facilitation of networking among industry participants through the establishment of a forum for the dialogue between the industry and local economic development officials and institutions. Such a forum focused primarily on the problems in linking the industrial actors within the cluster with the local community, may help the cohesion among different actors;
- Encouragement of mature companies to partner with and provide space for start-ups. This solution improve the commercial awareness of start-up companies and provide them additional "incubator" style space and

business mentoring, allowing large companies with a close relation with the most innovative products and processes.

In the mid 90's industry leaders in San Diego, gathered together with a strong commitment to create an association that would ensure growth and expansion opportunities and represent the industry's interests on a local, state and national level, biocom was founded in 1995 by the merger of the biomedical industry Council (BIC) and the San Diego Biocommerce Association. The organization was initially created to provide advocacy for industry on future industry growth. Over the last six years, the Biocom has grown into one of the largest and most acknowledged life science regional trade associations in the nation. Biocom currently operates for members in the areas of public policy advocacy, industry events and conferences, promotion of the industry, professional development programs, industry news and information, and most importantly, purchasing group and member discounts that substantially affect the bottom line of the companies' value chain.

Support to R&D Outsourcing Processes and Industrial Spin-offs

As a consequence of the biotech revolution that changed the way of research activities are carried out in the pharmaceutical sector, most big Pharmas tend to specialize in development and marketing, outsourcing the research phase. A broader and more complex set of biotech-based technologies and disciplines emerged, thus "eroding" the leadership in innovation of traditional large organizations in favor of the smaller DBFs. Moreover, innovations in the sector are significantly more rapid than in the past and it is much more difficult to keep in house activities up to speed in all fronts. Restructuring processes of large industries in recent years allowed the birth of many innovative biotech industries through spin-off mechanisms. Direct intervention in supporting and favoring this dynamic, which concern high value-added activities, should be implemented to enhance the generation of new industries

In order to support the creation of new industries through the R&D outsourcing processes, the following practice can be highlighted: favor of MBOs and industrial spin-offs from large corporations. Such mechanisms allow exploiting the industrial research, "re-vitalizing" the existing labs and infrastructures (otherwise dismissed) and maintaining a high level of employment.

The Bio valley, centered in the triangle Friburg, Strasbourg and Basel, developed recently under the joined effort of the governments of Germany, France and Switzerland.The idea to support the biotechnological sector was initially conceived by some private German actors at the end of the '80s. Nevertheless, up to 1996, the related initiatives were rather rare. In 1996 the cluster effectively started when the process of M&A leading to the creation of Novartis created nearly 3,000 qualified people unemployed in the life science sector. Based on a joint initiative of the regional and local governments, development agencies, universities and private companies the Bio valley Promotion Team was founded. The Bio valley Promotion Team had the main objective of supporting the creation of a biotech cluster, and primarily the spin-off process and the R&D outsourcing from Novartis. The annual budget of the Bio valley Promotion Team was around €1 million, initially from public funds and subsequently mixed (private and public). The project aimed at creating 400 new DBFs (mainly industrial and academic spin-off) and of 3,000 new jobs on a 10-year horizon. Till now, the initiative gained good results, facilitating the creation of 121 new biotech companies (from 1997 to 2000).

The birth of bio search Italia is heavy linked to the mentioned restructuring processes of large pharmaceutical companies. Up to 1995, it was a leading edge research centre of the Lepetit Group in Italy, focused on the discovery and production of antibiotics. In those years, the Lepetit Group was acquired by the Marion Merrelll Dow, which soon afterwards merged with Hoechst and created the Hoechst Marion Roussel, HMR (now Aventis). As a result of these M& Aaa processes, the Italian research centre became "Non-core" and was doomed to demission". The local management, however, decided in 1996 to start an independent company (Bio search Italia) and, through a management-Buy-out, "acquired" the centre (infrastructures, human resources, patents...) HMR favored the MBO with two years research contract of €14 million worth. Biosearch Italia also gained the access to public funds by the MIUR (Italian Ministry for the University and Research) for nearly €15 million.

In March 2003, Versicor and Biosearch Italia completed a merger to create an international company (Vicuron Pharmaceuticals) focused on anti-infective.

In 1994 Bayer established research centre in Milano (Italy) for the development of biochemical and bio molecular assays. At the end of nineties, Bayer changed its strategy and carried out a strong focus on pharmaceutical units, planning the closure of the other technology and service development units and the resource of the outsourcing. In Italy, Bayer favored the spin-off of Axxam from its centre in Milan, granting the new company a five year research contract of €29 million.

Mechanisms to Attract Key Managerial and Commercial People

Even if R&D actually represents the more activity of DBFs, running a successful company requires broader competences, particular regarding management and marketing. As regarding the scientific base, to attract key managerial and commercial) and commercial people strengthens its possibilities of success, making available in loci an experienced management team for new and existing companies.

In order to better attract key managerial and commercial people, the following practices can be highlighted:

- Promotion of new business ideas to the community. A rapid and broad delivery of the relevant information concerning the main areas of activities and of the most promising business ideas may convince top-managers to move.
- Support of creation of specialized consultancy and service companies. Lawyers, accountants, patent agents, recruitment agents and IT support, dedicated to the biotech, are needed in order to create a "favorable environment" for the growth of the industrial base and to make the geographic area "suitable" for top-level managers.

Availability of Specific Service and Infrastructures

In order to facilitate the development of an adequate industrial base within the cluster particularly, dedicated infrastructure is needed. Among the usual services the following can be highlighted: (i) incubators, providing spaces and shared service facilities (secretary as well as wet labs) for early stage start-ups; (ii) science parks, providing analogous infrastructures and services, house and support biotech companies emerging from incubators as well as later stage companies; and (iii) hospitals and clinics, supporting the clinical development phases for biotech drugs and diagnostic technologies.

In order to enhance the availability of specific services and infrastructures, the following actions can be undertaken:

- Creation of public biotech-dedicated infrastructures, through direct investments from t he local or central government, Public incubators and science parks, as no profit organizations, represent an effective way for many small start-ups to overcome the lack of capital, reducing initial investment needs;
- Promotion of the development of pilot manufacturing and contract manufacturing facilities (bio-processing, fermentation, and cell culture), even "inside" large existing companies. Indeed, it is very difficult, in capital

shortage, for new companies to build their own manufacturing facilities;

- To support of the creation of corporate incubators. This practice goes further than the previous one. It represents, for large companies, an effective way to "saturate" their plants and infrastructures, to gain alternative revenues and to engage close relations with innovative start-ups. At the same time, start-ups gain the access to start of the art facilities to develop their scientific and business ideas.

Moreover, the presence of specialized service providers is needed too. Lawyers, patent assistants as well as plant engineers in the geographical area of the cluster may facilitate the establishment of new industrial sites, reducing searching costs for new companies.

Leading Product Technologies

Early biologics were protein-based drugs produced by splicing genes into bacteria (*e.g.*, recombinant, insulin, human growth hormone, and erythropoietin). Biotech companies now devote a larger share of their product portfolios to newer technologies, such as antisense, immune-based therapy, and monoclonal antibodies (mAbs).

Almost a third-90-of the products in the biotech pipeline are therapeutic vaccines, many of which promise to jump-start the immune system to fight AIDS and several cancers. The pipeline also is brimming with mAbs, developed from a lab-made version of a naturally occurring protein that binds to and neutralizes foreign invaders. Recently, the first recombinant mAb to stop angiogenesis, the formation of new blood vessels that feed tumors, was approved for the treatment of meta static colorectal cancer.

Other promising products in the pipeline include antisense drugs-designed to interfere with the communication process that signals a cell to produce an unwanted protein. Fourteen antisense products are in development to treat AIDS, certain cancers, Crohn's disease, and heart disease. Ten interferon protein products, which interfere with cells' ability to reproduce, are in development. If approved, they will be used to treat multiple sclerosis, osteoporosis, hairy cell leukemia, chronic granumlomatous disease, genital warts, and several other diseases.

IZB is the largest incubator center in the biotech of Munich. Launched in 1995, the initiative was supported by public actor at local and regional level. The total financing by the Bavarian government was about €45 million. Due also to the winning of the BioRegio initiative the centre became one of the most important actor for the support of first phases of start-up companies in the Munich area. The incubator offers logistics services and some shared machineries. Furthermore the structure hosts bio-m-AG which offers many other services (consultants, financing…) because of the great success of the initiative, and with the aim to differentiate the focus of the biotech, a second incubator focusing on green biotech was created. The first site currently hosts 23 enterprises, some of which passed the start-up phase with 2 companies going public in 2002. Seven companies are currently hosted in the new site.

Public Acceptance of Biotech Activities

Public acceptance of biotech activities refers to the positive "feeling" of the social community towards the sector. A favorable social environment may have a strong effect on clusters growth, "spurring on" the workforce and encouraging entrepreneurship in biotech activities. In order to enhance the public acceptance of biotechnology within the cluster, the following actions can be undertaken:

- Open debates about benefit and risks of biotechnology promoting mutual understanding between the general public and the scientific community. Social as well as ethical issues should be discussed, avoiding common "prejudices" and adopting a scientific perspective, thus disseminating a correct view of biotech possibilities and risks;

- Development of a program for the local biotechnology industry to interact with local high school and community college teachers and students regarding the job potential and opportunities within the cluster. The programme should also allow the development of curricula to meet the basic educational needs of the biotech industry.
- Development of seminar programmes and hand-on workshops at school and business sites, eventually encouraging younger employees to go out into the community "to spread the world" on biotechnology activities and career opportunities;
- Public promotion of companies and university lab activities, for example increasing the number of business Open Day, thus allowing the public to know what goes on in biotech labs.

In June 1998. the Swiss voted in favor of encouraging research in genetic engineering, demonstrating a diffused acceptance and acknowledgement of biotechnology. Moreover, a strong action to enhance the public acceptance of biotech activities is engaged by the Swiss government. As a service to the Swiss society, two agencies have been created to assess the impact of biotechnology applications, and to promote informed public debates and decision making. Both agencies participate in national and international networks. The Biosafety Research and assessment of Technology Impacts (BATS) focuses in the scientific safety assessment of transgenic plants and foods derived from genetically modified organisms (GMOs). It also participates in national and international projects on the assessment of social and ethical aspects of technology impact. The Swiss agency for biotechnology Information and Communication (BICS) was instead created to fulfill the public's need for information about this rapidly changing sector-k with a specialized library and an information service on biotechnology available online. A website (www.bioweb.org) created in 2000, provides a huge variety of biotech related contents (currently 684 articles, 917 news items, 1,312 link with descriptions, library catalogue with 3,981 entries and an integrated glossary with 425 terms,) in English, French and German.

Another important program in supporting biotech acceptance, an generally technological and scientific culture's improvement among people, is ": Science et Cite". It is a festival sponsored by the Swiss Government in university cities to promote mutual understanding between the general public and the scientific community. The festival's program comprehends different activities (monthly discussion forums in different cities' Café, a permanent roundtable currently on environmental problems, a newsletter and forum service regarding the theme of stem cell research) and is supported also by big national biotech firms, for example Serono.

The Gene Cafes are quarterly meetings base on the model of "Philosophy Cafes" and take place in different cafes, restaurants near Envy. They offer the public an opportunity to meet and ask questions to the scientists and industrialists working on campus. At each meeting, Pierre Tambourin, Chief Executive of Genopole, hosts one or more speakers. Among the recently debated theme: the correct use of genetics, the benefits and risks of therapeutic clones, the biotechnology impact on the environment

Future Challenges

Over the next 30 years, the platoons of developing countries will continue to grow at a rate of 1.6% per year and the expected income growth will further increase the demand for maize and wheat. Overall, the demand for wheat and maize is expected to grow at an annual rate of 2.2% over the next 30 years. By the year 2020, 67% of the world's wheat consumption and 55% of the world's maize consumption will occur in developing countries. Projections indicate that they will need to import about 122 million metric tonnes of wheat and 41 million metric tonnes of maize annually. Moreover, nearly all the necessary increases in maize and wheat production in the developing world will have to emanate from increased yields since, with the exception of very few countries, the opportunities to increase the area under cultivation are limited. Agricultural research must continue to play a critical role in developing new technology to further increase the productivity of maize and wheat cropping. Its dominant challenge for the future is in deciding how to proceed beyond the interim stage of development that the present formative structure of projects represents.

Considerable skill will be required to sustain financial security without distorting the mission and the work programmes designed to achieve it.

The management structure is very important for a project of this size and sophistication. The coordinator of BioSapiens has responsibility for establishing the network and achieving

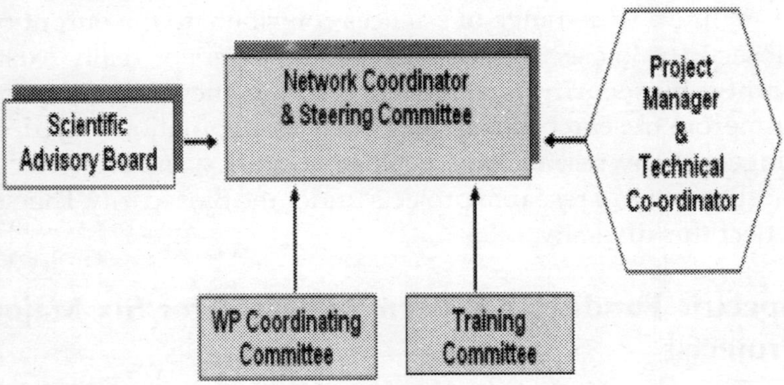

Fig. 12

its objectives, including project reporting and accounting. A project manager, whose role is to provide day-to-day support and implementation of all management and organisational tasks, supports the coordinator. A Steering Committee, chaired by the Coordinator, oversees the work of the network and assists the coordinator in the management of BioSapiens. The Steering Committee meets at least quarterly.

In addition to the Steering Committee, a Work Package Coordinating Committee and a Training Committee oversee the progress of the work packages and training activities respectively.

Bio Security

Bio security (often also known as quarantine) is the outcome of an eclectic mix of sciences and regulatory tools that together, contribute to the exclusion and suppression of unwanted organisms.

Current biosecurity systems are increasingly under pressure from increasing imports (up 76% in the last decade) and tourism (up 93% in the last decade). The obvious response to these increasing risks has been to intensify existing operational systems; however, this approach eventually leads to diminishing returns. Inevitably, new scientific approaches and technologies must be employed to keep pace with growing bio security threats. This calls for research and development.

While a wide range of sciences contribute to the output of biosecurity, 'biosecurity science' per se does not really exist. Rather, biosecurity is achieved through the integration of numerous research disciplines such as population genetics, biogeography, risk analysis, ecology, sociology, and economics. Bio-Protection's research projects under the Biosecurity Theme reflect this diversity.

Specific Funding has been Allocated for Six Major Projects:

The use of neural network computing to model and predict species invasiveness

The development of molecular-based identification systems for invasive species

Predicting weed naturalisation, spread and decline (Details to come)

Exploiting the host finding behaviour of thrips: underlying principles

Emerging technologies for border diagnostics: Pest identification using genetic and stable isotope signatures

Intelligent Systems for Biosecurity

Closely related to the BioProtection-funded research, are a number of closely integrated FRST-funded programmes. This includes a very large joint programme involving AgResearch, Crop & Food Research, HortResearch and MAF which is addressing border biosecurity for the agricultural, horticultural and cropping sectors. Similarly, a more specific programme has been established to research the application of sensor technologies designed to strengthen container biosecurity (known as SniffertechTM). It is also important to recognise that many existing pest management programmes throughout New Zealand have strong biosecurity components and these again have links with Bio-Protection. This particularly applies to those programmes researching insect pheromones (attractant chemicals), which have immediate utility to bio security.

Coupled with the above, significant funding has also been made available to Bio-Protection for advanced climate simulation facilities. One of Bio-Protection's immediate challenges is to achieve integration of the widespread research being done. To make progress Bio-Protection has begun an on-going process of co-operation with the Ministry of Agriculture and Forestry's Biosecurity agencies, its partner institutions and other agencies throughout world. This approach has been publicized through a number of presentations to stakeholders

Biocontrol

Biocontrol is the use of living organisms to control or reduce populations of unwanted (pest) organisms. Biocontrol of pest animals, weeds and diseases can be elegant, self-sustaining, non-polluting and inexpensive. The attraction of biocontrol, which purports to practice "clean and green" agriculture, is that it can complement or substitute for chemical pesticides. The use of pesticides is becoming an increasingly displeased issue. Concerns about adverse affects to human health and the environment have decreased the public's acceptance of pesticide use. Often the very large areas involved and the relatively low value of the crop per hectare mean that pesticide use is uneconomic, and our better understanding of ecology has highlighted a range of inherent difficulties in targeting pesticides to pests *e.g.* some key pasture pests have feeding sites that are not exposed.

Two general strategies for biocontrol exist. The classical (inoculative) approach involves establishing populations of a pest's natural enemies. This involves importing new bio control species from overseas and manipulating the agro-ecological system to favour the bio control agent (a practice termed 'ecological engineering'). The classical strategy leads to a self-perpetuating regulation of pest populations rather than total eradication, and has thus become a measure by which to protect and restore natural pest suppression, not just as non-polluting pesticide replacements.

The second bio control strategy, the inundative approach, utilizes bio control organisms in a manner that mimics the use of chemical pesticides. Bio control agents (usually bacteria or fungi) are formulated and applied in massive doses to simulate epidemic conditions, after which the bio control agent population decreases to a low level or dies out. A key benefit of this strategy is that, through formulation and application technologies, bio control agents may be applied in a well-targeted manner. Because inundate bio controls ('biopesticides') are formulated and applied annually, it is possible to market them as commercial products.